SYSTEM ADMINISTRATION

Pankaj Sharma
Department of Information Technology
ABES, Ghaziabad

S.K. KATARIA & SONS®
Publisher of Engineering Books
4885/109, Prakash Mahal, Dr. Subhash Bhargav Lane,
Opposite Delhi Medical Association, Daryaganj, New Delhi–110002 (INDIA)
Phone: +91-11-43551243, *Mobile:* +91-9871775858, 7678425209
e-mail: katariabook@gmail.com, katariabooks@yahoo.com
Website: www.skkatariaandsons.com
(e-books, subscriptions and bulk buying also available on our website)

Published by:

S.K. Kataria & Sons®
4885/109, Prakash Mahal, Dr. Subhash Bhargav Lane,
Opposite Delhi Medical Association, Daryaganj, New Delhi-110002 (INDIA)
Phone: +91-11-43551243; *Mobile:* +91-9871775858, 7678425209
e-mail: katariabook@gmail.com, katariabooks@yahoo.com
Website: www.skkatariaandsons.com

First Edition: July, 2012, Reprint: 2026

Price: ₹ 395/-

Laser Type Setting:
PLUS COMPUTERS, Shahdara, Delhi

Printed at:
New A.S. Printing Press, Delhi

Preface.....

System Administration is a core subject for under gradute students and postgraduate students in computer science Engineering Information Technology streams and master of computer application (MCA). This book is primarily intended to serve as a text book in accordance with the syllabus of system administration offered by various universities in India as well as abroad.

In this book, a significant effort has been made to find simple ways to develop theoretical as well as Practical concepts of system administration. Neat and clean diagrams has been used for explanation.

This book is divided into nine chapters :

The organisation of these chapters are :

Chapter-1 : Describes the Role of Linux System Administrator.

Chapter-2 : Describes Various Administration Tools and their Accessibility

Chapter-3 : Introduces Various Methods for Managing user Accounts

Chapter-4 : Introduces Various Methods for Managing UNIX File Systems

Chapter-5 : Describes How to Configure Linux Kernal

Chapter-6 : Describe How to Configure the TCP/IP Networking

Chapter-7 : Introduces the Concept of TCP/IP Firewalls

Chapter-8 : Describes Various Concepts of Network Information System

Chapter-9 : Describes the Various Concepts of Active Directory and IDAP

Pankaj Sharma

Acknowledgement.....

I get this opportunity to express my gratitude and thanks to the respected Director (administration) of ABES Engg. College, **Dr. S.K. Kansal** for his valuable technical suggestion and constant encouragement, without which this book would not have to come into existence.

I express kind gratitude of **Prof. R.G. Mendiratta**, Director (Academic) ABES Engg. College, Ghaziabad

I am grateful to **Prof. V.K. Srivastava,** Director of ABES IT, Ghaziabad, for helping me through his insightful suggestions and construction inputs.

I wish to thank **Prof. Pradeep Sharma**, Head of Department of Information Technology, Department of IEC, Greater Noida, for their constant co-operation, directly inspiration and encourangement.

I like to remember the motivation initiated by **Prof. D. Pandey** (HOD CSE) and Prof. Sunil Sharma (AHODCSE) of ABES, Engg. College. I am also thankful to Prof. K.P. Jayant (HODMCA) for his inspiration.

I like to remember the motivation initiated by My Respected Father in Law **Mr. M.D. Punia** (Lead District Manager of PNB, Karnal (Haryana).

I thank to my elder brother **Mr. Sanjay Sharma** (KIET), Ghaziabad for inspiration and encouragement.

I thank to **Mr. Ashwim Perti,** Sr. Lecturer, Gzb. for inspiration and encouragement.

I am grateful to my younger Brother **Sandeep** and daughter **Prisha** for inspiring me for this project.

I deeply express my heartful thanks to the **Mr. Sanjeev Kataria** for **M/s S.K. Kataria & Sons** for publishing this book in such a beautiful getup and well in time.

Lastly I am grateful to Mr. Manoj Tyagi and his staf from Plus Computer's for their hard work.

Pankaj Sharma

Dedicated This book to

My Parents, Wife Priyanka

and

Daughter Prisha (Guddu) & Pihu

Contents.....

CHAPTER 1

THE LINUX SYSTEM ADMINISTRATOR

Chapter Outline

- Introduction
- The Linux System Administrator
- System and Network Administration
- Installing and Configuring Servers
- Installing and Configuring
- Creating and Maintaining
- Backing Up and Restoring Files
- Monitoring and Tuning Performance
- Configuring a Secure System
- Using Tools to Monitor Security

1.0 INTRODUCTION

LINUX IS A MULTIUSER, multitasking operating system from the ground up. In this regard the system administrator has flexibility — and responsibility — far beyond those of other operating systems. Red Hat has employed innovations that extend these duties even for the experienced Linux user. In this chapter, we briefly look at those responsibilities, which we'll cover in more detail in later chapters.

1.1 The Linux System Administrator

Using Linux involves much more than merely sitting down and turning on the machine. Often you hear talk of a "steep learning curve" but that discouraging phrase can be misleading. Instead, Linux is quite different from the most popular commercial operating systems in a number of ways. While it is no more difficult to learn than other operating systems, it is likely to seem very strange even to the experienced administrator of other systems. In addition, the sophistication of a number of parts of the Red Hat distribution has increased by an order of magnitude, so even an experienced Linux administrator is likely to find much that is new and unfamiliar. Fortunately, there are new tools designed to make system administration easier than ever before.

1.2 SYSTEM AND NETWORK ADMINISTRATION

Make no mistake: Every computer in the world has a system administrator. It may be — and probably is — true that the majority of system administrators are those who decided what software and peripherals were bundled with the machine when it was shipped. That status quo remains because the majority of users who acquire computers for use as appliances probably do little to change the default values. But the minute a user decides on a different wallpaper image or adds an application that was acquired apart from the machine itself, he or she has taken on the role of system administration.

The highfalutin title of system administrator brings with it some responsibilities. No one whose computer is connected to the Internet, for instance, has been immune to the effects of poorly administered systems, as demonstrated by the Distributed Denial of Service (DDoS) and e-mail macro virus attacks that have shaken the online world in recent years. The scope of these acts of computer vandalism (in some cases, computer larceny) would have been greatly reduced if system administrators had a better understanding of their duties.

Linux system administrators are likely to understand the necessity of active system administration more than those who run whatever came on the computer, assuming that things came properly configured from the factory. The user or enterprise that decides on Linux has decided, too, to assume the control that Linux offers, and the responsibilities that this entails.

By its very nature as a modern, multiuser operating system, Linux requires a degree of administration greater than that of less robust, home-market systems. This means that even if you use just a single machine connected to the Internet by a dial-up modem — or not even connected at all — you have the benefits of the same system employed by some of the largest businesses in the world, and will do many of the same things that IT professionals employed by those companies are paid to do. Administering your system does involve a degree of learning but it also means that in setting up and configuring your own system you gain skills and understanding that raise you above mere "computer user" status. The Linux system administrator does not achieve that mantle by purchasing a computer but by taking full control of what the computer does and how it does it.

You may end up configuring a small home or small office network of two or more machines, perhaps including ones that are not running Linux. You may be responsible for a business network of dozens of machines. The nature of system administration in Linux is surprisingly constant, no matter how large or small your installation. It merely involves enabling and configuring features you already have available.

By definition, the Linux system administrator is the person who has "root" access, which is to say the one who is the system's "super user" (or root user). A standard Linux user is limited to whatever he or she can do with the underlying engine of the system. But the root user has unfettered access to everything — all user accounts, their home directories, and the files therein; all system configurations; and all files on the system. A certain body of thought says that no one should ever log in as "root," because system administration tasks can be performed more easily and safely through other, more specific means, which we discuss in due course. Because the system administrator has full system privileges, your first duty is to know what you're doing, lest you break something.

The word duty implies a degree of drudgery; in fact, it's a manifestation of the tremendous flexibility of the system measured against the responsibility to run a tight organization. These duties do not so much constrain you, the system administrator, as free you to match the job to the task. Let's take a brief look at them.

1.2.1 Installing and Configuring Servers

When you hear the word server to describe a computer, you probably think of a computer that offers some type of service to clients. The server may provide file or printer sharing,

File Transfer Protocol (FTP) or Web access, or e-mail processing tasks. Don't think of a server as a standalone workstation; think of it as a computer that specifically performs these services for many users.

In the Linux world, the word server has a broader meaning than what you might be used to. For instance, the standard Red Hat graphical user interface (GUI) requires a graphical layer called XFree86. This is a server. It runs even on a stand- alone machine with one user account. It must be configured. (Fortunately, Red Hat has made this a simple and painless part of installation on all but the most obscure combinations of video card and monitor; gone are the days of anguish as you configure a graphical desktop.)

Likewise, printing in Linux takes place only after you configure a print server. Again, this has become so easy as to be nearly trivial.

In certain areas the client-server nomenclature can be confusing, though. While you cannot have a graphical desktop without a server, you can have Web access without a Web server, FTP access without running an FTP server, and e-mail capabilities without ever starting a mail server. You may well want to use these servers, all of which are included in Red Hat; then again, maybe not. Whenever a server is connected to other machines outside your physical control, there are security implications to consider. You want your users to have easy access to the things they need but you don't want to open up the system you're administering to the whole wide world.

Whenever a server is connected to machines outside your physical control, security issues arise. You want users to have easy access to the things they need but you don't want to open up the system you're administering to the whole wide world.

Linux distributions used to ship with all imaginable servers turned on by default. Just installing the operating system on the computer would install and configure —with default parameters — all the services available with the distribution. This was a reflection of an earlier, more innocent era in computing when people did not consider vandalizing other people's machines to be good sportsmanship.

Unfortunately, the realities of this modern, more dangerous world dictate that all but the most essential servers remain turned off unless specifically enabled and configured. This duty falls to the system administrator. You need to know exactly which servers you need and how to employ them, and to be aware that it is bad practice and a potential security nightmare to enable services that the system isn't using and doesn't need. Fortunately, the following pages show you how to carry out this aspect of system administration easily and efficiently.

1.2.2 Installing and Configuring

Application Software

Although it is possible for individual users to install some applications in their home directories — drive space set aside for their own files and customizations — these applications are not available to other users without the intervention of the system administrator. Besides, if an application is to be used by more than one user, it probably needs to be installed higher up in the Linux file hierarchy, which is a job that only the system administrator can perform. (The administrator can even decide which users may use which applications by creating a "group" for that application and enrolling individual users in that group.)

New software packages might be installed in/opt if they are likely to be upgraded separately from the Red Hat distribution itself. Doing this makes it simple to retain the old version until you are certain that the new version works and meets your expectations. Some packages may need to go in /usr/local or even /usr if they are upgrades of packages installed as part of Red Hat. (For instance, there are sometimes security upgrades of existing packages.) The location of the installation usually matters only if you compile the application from source code; if you use a Red Hat Package Manager (RPM) application package, it automatically goes where it should.

Configuration and customization of applications is to some extent at the user's discretion, but not entirely. "Skeleton" configurations — administrator-determined default configurations — set the baseline for user employment of applications. If there are particular forms, for example, that are used throughout an enterprise, the system administrator would set them up or at least make them available by adding them to the skeleton configuration. The same applies to configuring user desktops and in even deciding what applications should appear on user desktop menus. For instance, your company may not want to grant users access to the games that ship with modern Linux desktops. You may also want to add menu items for newly installed or custom applications. The system administrator brings all this to pass.

1.2.3 Creating and Maintaining

User Accounts

Not just anyone can show up and log on to a Linux machine. An account must be created for each user and — you guessed it — no one but the system administrator can do this. That's simple enough.

But there's more. It involves decisions that either you or your company must make. You might want to let users select their own passwords, which would no doubt make them easier to remember but which probably would be easier for a malefactor to crack. You might want to assign passwords, which is more secure in theory but increases the likelihood that users will write them down on a conveniently located scrap of paper — a risk if many people have access to the area where the machine(s) is located. You might decide that users must change their passwords periodically — something you can configure Red Hat Enterprise Linux to prompt users about.

What happens to old accounts? Suppose someone leaves the company. You probably don't want him or her to gain access to the company network, but you also don't want to delete the account wholesale, only to discover later that essential data resided nowhere else.

To what may specific users have access? It might be that there are aspects of your business that make Web access desirable, but you don't want everyone spending their working hours surfing the Web. If your system is at home, you may wish to limit your children's access to certain Web sites.

These and other issues are part of the system administrator's duties in managing user accounts. Whether the administrator or his or her employer establishes policies governing accounts, these policies should be delineated— preferably in writing for a company — for the protection of all concerned.

1.2.4 Backing Up and Restoring Files

Until computer equipment becomes infallible, until people lose the desire to harm others' property, and — truth be told — until system administrators become perfect, there is considerable need to back up important files so that the system can be up and running again with minimal disruption in the event of hardware, security, or administration failure.

Only the system administrator may do this. (Because of its built-in security features, Linux doesn't allow users even to back up their own files to removable disks.)

It's not enough to know that performing backups is your job. You need to formulate a strategy for making sure your system is not vulnerable to catastrophic disruption. This is not always obvious. If you have a high-capacity tape drive and several good sets of restore disks, you might make a full system backup every few days. If you are managing a system with scores of users, you might find it more sensible to back up user accounts and system configuration files, figuring that reinstallation from the distribution CDs would be quicker and easier than getting the basics off a tape archive. (Don't forget about applications you install separately from your Red Hat distribution, especially those involving heavy customization.)

Once you decide what to back up, you need to decide how frequently to perform backups, whether to maintain a series of incremental backups — adding only files that have changed since the last backup — or multiple full backups, and when these backups should be performed. Do you trust an automated, unattended process? If you help determine which equipment to use, do you go with a redundant array of independent disks (RAID), which is to say multiple hard drives all containing the same data as insurance against the failure of any one of them, in addition to other backup systems? (A RAID is not enough because hard drive failure is not the only means by which a system can be brought to a halt.)

You don't want to become complacent or foster a lackadaisical attitude among users. Part of your strategy should be to maintain perfect backups without ever needing to resort to them. This means encouraging users to keep multiple copies of their important files in their home directories so that you won't be asked to mount a backup to restore a file that a user corrupted. (If your system is a standalone one then, as your own system administrator, you should make a habit of backing up your configuration and other important files.)

Restoring files from your backup media is no less important than backing them up in the first place. Be certain you can restore your files if the need arises by testing your restore process at least once during a noncritical time.

Chances are good that even if you work for a company, you'll be the one making these decisions. Your boss just wants a system that runs perfectly, all the time. Backing up is only part of the story, however. You need to formulate a plan for bringing the system back up after a failure. A system failure could be caused by any number of problems, either related to hardware or software (application, system configuration) trouble, and could range from a minor inconvenience to complete shutdown.

Hardware failures caused by improper configuration can be corrected by properly configuring the device. Sometimes hardware failures are caused by the device itself, which typically requires replacing the device. Software failures caused by improperly configured system files are usually corrected by properly configuring those files. An application can cause the system to fail for many reasons and may require a lot of research on the part of the administrator to find the root of the problem.

If you are the administrator of servers and workstations for a business, you should have a disaster recovery plan in place. Such a plan takes into account the type of data and services provided and how much fault tolerance your systems require — that is, how long your systems could be down and what effect that would have on your company's ability to conduct business. If you require 100 percent fault tolerance, meaning your systems must be online 24/7, then disaster recovery is unnecessary as your systems never go down and there is no disaster from which to recover. Most organizations, though, cannot afford such a high level of fault tolerance; they are willing to accept less stringent standards.

Based on the level of fault tolerance you require, your disaster recovery plan should list as many possible failures as you can anticipate and detail the steps required to restore your systems. In chapter 2, we describe fault tolerance and disaster recovery in more detail.

Backing up is only part of the story. You need to formulate a disaster recovery plan to bring your system back up in the event of a failure.

1.2.5 Monitoring and Tuning Performance

The default installation of Red Hat Enterprise Linux goes a long way toward capitalizing on existing system resources. There is no "one size fits all" configuration, however. Linux is infinitely configurable, or close to it.

On a modern standalone system, Linux runs pretty quickly. If it doesn't, there's something wrong — something the system administrator can fix. Still, you might want to squeeze one last little bit of performance out of your hardware — or a number of people might be using the same file server, mail server, or other shared machine, in which case seemingly small improvements in system performance add up.

System tuning is an ongoing process aided by a variety of diagnostic and monitoring tools. Some performance decisions are made at installation time, while others are added or tweaked later. A good example is the use of the hdparm utility, which can increase throughput in IDE drives considerably; but for some high-speed modes a check of system logs shows that faulty or inexpensive cables can, in combination with hdparm, produce an enormity of nondestructive but system-slowing errors.

Proper monitoring allows you to detect a misbehaving application that consumes more resources than it should or fails to exit completely upon closing. Through the use of system performance tools you can determine when hardware —such as memory, added storage, or even something as elaborate as a hardware RAID — should be upgraded for more cost-effective use of a machine in the enterprise or for complicated computational tasks such as three-dimensional rendering.

Possibly most important, careful system monitoring and diagnostic practices give you an early heads-up when a system component is showing early signs of failure, so that you can minimize any potential downtime. Combined with the resources for determining which components are best supported by Red Hat Enterprise Linux, performance monitoring can result in replacement components which are far more robust and efficient in some cases.

In any case, careful system monitoring plus wise use of the built-in configurability of Linux allows you to squeeze the best possible performance from your existing equipment, from customizing video drivers to applying special kernel patches or simply turning off unneeded services to free memory and processor cycles.

1.2.6 Configuring a Secure System

If there is a common thread in Linux system administration, it is the security of the computer and data integrity.

What does this mean? Just about everything. The system administrator's task, first and foremost, is to make certain that no data on the machine or network are likely to become corrupted, whether by hardware or power failure, by misconfiguration or user error (to the extent that the latter can be avoided), or by malicious or inadvertent intrusion from elsewhere. It means doing all the tasks described throughout this chapter, and doing them well, with a full understanding of their implications.

No one involved in computing has failed to hear of the succession of increasingly serious attacks on machines connected to the Internet. For the most part, these attacks have not targeted Linux systems. That doesn't mean Linux systems have been entirely immune, either to direct attack or to the effects of attacks on machines running other operating systems. In one Distributed Denial of Service (DDoS) attack aimed at several major online companies, for instance, many "zombie" machines — those that had been exploited so that the vandals could employ thousands of machines instead of just a few — were running Linux that had not been patched to guard against a well-known security flaw. In the various "Code Red" attacks during the summer of 2001, Linux machines themselves were invulnerable, but the huge amount of traffic generated by this "worm" infection nevertheless prevented many Linux machines from accomplishing much Web-based work for several weeks, so fierce was the storm raging across the Internet. And few e-mail users have been immune from receiving at least some "SirCam" messages — nonsensical messages from strangers with randomly selected files attached from their machines. While this infection did not corrupt Linux machines per se, as it did those running MS Windows, anyone on a dial-up Internet connection who had to endure downloading several megabytes of infected mail each day would scarcely describe himself or herself as unaffected by the attack.

Depending on how a Linux machine is connected, and to what, the sensitivity of the data it contains and the uses to which it is put, security can be as simple as turning off unneeded services, monitoring the Red Hat security mailing list to make sure that all security advisories are followed, and otherwise engaging in good computing practices to make sure the system runs robustly. It's almost a full-time job involving levels of security permissions within the system and systems to which it is connected; elaborate firewalls to protect not just Linux machines but machines that, through their use of non-Linux software, are far more vulnerable; and physical security— making sure no one steals the machine itself!

For any machine connected to another machine, security means hardening against attacks and making certain no one else uses your machine as a platform for launching attacks against others. If you run Web, FTP, or mail servers, it means giving access to only those who are entitled to it, while locking out everyone else. It means making sure that passwords are not easily guessed and not made available to unauthorized persons. It means that disgruntled former employees no longer have access to the system and that no unauthorized person may copy files from your machines.

Security is an ongoing process. The only really secure computer is one that contains no data, is unplugged from networks and power supplies, has no keyboard attached, and resides in a locked vault. While that is theoretically true, it implies that security diminishes the usefulness of the machine. Your job as system administrator is to strike the right balance between maximum utility and maximum safety, all the while bearing in mind that confidence in a secure machine today means nothing about the machine's security tomorrow.

In the chapters that follow, you learn about the many tools that Red Hat provides to help you guard against intrusion, even to help you prevent intrusion into non-Linux machines that may reside on your network. Linux is designed from the beginning with security in mind. In all your tasks you should maintain that same security awareness.

Your job as system administrator is to strike the right balance between maximum utility and maximum safety, all the while bearing in mind that confidence in a secure machine today means nothing about the machine's security tomorrow.

1.2.7 Using Tools to Monitor Security

People who, for purposes of larceny or to amuse themselves, like to break into computers

— they're called "crackers" — are a clever bunch. If there is a vulnerability in a system, they will find it. Fortunately, the Linux development community is quick to find potential exploits and to create ways of slamming the door shut before crackers can enter. Fortunately, too, Red Hat is diligent in making available new, patched versions of packages in which potential exploits have been found. Your first and best security tool, therefore, is making sure that whenever a security advisory is issued, you download and install the repaired package. This line of defense can be annoying but it is nothing compared to rebuilding a compromized system.

As good as the bug trackers are, sometimes their job is reactive. Preventing the use of your machine for nefarious purposes and guarding against intrusion are, in the end, your responsibility alone. Red Hat equips you with tools to detect and deal with unauthorized access of many kinds. As this book unfolds, you'll learn how to install and configure these tools and how to make sense of the warnings they provide. Pay

SUMMARY

1. Bell Labs had to withdraw themselves from the Multics project. Ken Thompson then started working on a simpler project and he named it UNIX.
2. This version of UNIX was rewritten in the year 1973. The source code of UNIX operating system was rewritten in C language by Dennis Ritchie, the inventor of C.
3. In order to make UNIX popular among users, AT & T came up with a unique marketing strategy. They started distributing source copies of UNIX to different universities at a very nominal price.
4. Thompson and Ritchie described the UNIX System and got it published in a newspaper named Communications of the ACM. This helped in increasing the acceptance level of the UNIX system even more.
5. The UNIX system was installed at around 500 different sites. UNIX system found its major contribution in the telephone companies, providing a good environment for program development, network transaction services and real time services.
6. A large number of institutions and universities were provided licenses of UNIX system. In the year 1977, the UNIX system was first ported from a PDP to a non-PDP machine.
7. The hardware of a UNIX system is present in the centre of the diagram. It provides the basic services such as memory management, processor execution level, etc. to the operating system.
8. The UNIX system seems to be organised as a set of layers. The system Kernel is also called the operating system.
9. The user programs are independent of the hardware on which they are running. Therefore, it becomes very easy to run programs on UNIX system running on different hardware if the programs do make special assumptions.
10. UNIX offers an excellent environment for developing new software. It provides a variety of tools ranging from editing a program to maintenance of software. It exploits the power of hardware to the maximum extent of effectiveness and efficiency.

SHORT QUESTIONS WITH ANSWER

1. Use ps to search for the init process by name.

Ans.
```
root@rhel53 ~# ps -C init
PID TTY         TIME CMD
1 ?             00:00:04 init
```

2. What is the process id of the init process ?

Ans. 1

3. Use the who am i command to determine your terminal name.

Ans.
```
root@rhel53 -# who am i
paul pts/0        2010-04-12 17:44 (192.168.1.38)
```

4. Using your terminal name from above, use ps to find all processes associated with your terminal.

Ans.
```
oot©rhel53 ~tf ps fax | grep pts/0
2941    ?       S     0:00     \_ sshd: paul®pts/0
2942    pts/0   Ss    0:00        \_ -bash
2973    pts/0   S     0:00           \_ su -
2973    pts/0   S     0:00              \_ -bash
3808    pts/0   R*    0:00              \_ ps fax
3809    pts/0   R+    0:00              \_ grep pts/O
```

or also

```
root@rhel53 -#       ps -ef | grep pts/o
paul    2941    2939  0  17:44  ?      00:00  sshd: paul®pts/0
paul    2942    2941  0  17:44  ptE/0  00:00  -bash
root    2972    2942  0  17:45  pts/0  00:00  su -
root    2973    2972  0  17:45  ptE/0  00:00  -bash
root    3816    2973  0  21:25  pts/0  00:00  ps -ef
root    3817    2973  0  21:25  pts/0  00:00  grep pts/0
```

5. What is the process id of your shell ?

Ans. `2973 in the screenshot above, probably different for you`

echo $$ should display same number as the one you found

6. What is the parent process id of your shell ?

Ans. `2972 in the screenshot above, probably different for you`

in this example the PPID is from the **su** - command, but when inside gnome then for example gnome-terminal can be the parent process

EXERCISE

1. What is meant by a multi-user and multi-tasking operating system?
2. What is the minimum space required of UNIX?
3. Who is the developer of UNIX.
4. In which year was first version released.
5. What are various Features of UNIX?
6. What makes UNIX portable and secure.
7. Can Unix be used as Network Operating System?

❐❐❐

CHAPTER 2

ADMINISTRATION TOOLS AND PROCESSES

2.0 INTRODUCTION

Simply stated, UNIX is an operating system. It is by and large the most popular operating system existing today. The features and flexibility of UNIX is so immense that it has become a standard for great many operating systems. It is multi-user system, which means that more than one user can work at the same computer system at the same time. UNIX also supports multi-tasking. Multi-tasking means that more than one program can be made to run at the same time. For example, you can initiate a program and leave it by itself to go on and in the meantime you can work on some other program. Multi-tasking and multi-user are the two most important characteristics of UNIX, which have helped it gain widespread acceptance among a large variety of users.

The UNIX OS files consume 40 MB of the 80 MB disk space. Another 10-20 MB of disk space is eaten up as swap space. This swap space is used at that point of time when UNIX falls short of memory. So, the contents that are not immediately required are stored in the swap space. Any time when the program needs these contents, they are read from the swap space.

Chapter Outline

- Introduction
- Linux System
- Overview of permissions
- Processes
- Process status
- Killing Process
- Process priority
- Linux Kernel
- Starting and Shut down Peripherals
- BIOS
- Console
- The scheduler
- The Init Program
- Runlevels
- The inittab file
- Run-levels and Run level scripts

2.1 LINUX SYSTEM ADMINISTRATION TOOLS

There are two best admin tools for Linux. They are Linuxconf and Webmin.

The Linux System administration tools are:

- Linuxconf Linuxconf is a extremely large project. It is tremendously powerful and has large number of features. Linuxconf is the BEST KNOWN system admin tool for Linux!! It is available in all distros like Redhat, Debian, Suse, and others. The Linuxconf is the equivalent of SAM of HPUX and SMITTY of AIX. It runs on console window and as well as in GUI in X-Window. On Debian Linux to download and install just do 'apt-get linuxconf'.

The Webmin is the most powerful and very popular system admin tool for Linux. Webmin is so popular and it was awarded the best System Administration tool for Solaris, BSD, Linux, HPUX, AIX, SCO and others. Webmin had been winning awards year after year for the past 5 years by many organizations, conferences and editors. Webmin can also be used for admin of other OSes like AIX, Solaris, HPUX, IRIX, BSD, SCO Unix, OSF, Darwin, Apple Macintosh Mac OS X and Cygwin. The best supported systems at the moment are Solaris, Linux (Redhat in particular) and FreeBSD. Webmin is written in Perl and Perl is universal like "C" and unlike "C" it is scripting language. Perl is "cousin-brother" of "C" language. But major drawback is that you need a X-Window system, Apache Webserver and Perl installed and running before using Webmin. And here is where Linuxconf fills the gap—you do not need X-Window and others to run Linuxconf. Linuxconf can run even in dumb text-terminal console with ncurses.

2.2 OVERVIEW OF PERMISSIONS

2.2.1 Permissions and Ownership Why?

If you can't access some of the files on your very own Linux system, it's usually because of misconfigured file access permissions. If you are the only user on your Linux box, you may be wondering what's the point of having all these permissions (or lack thereof) that restrict your access to your very own penguin OS. However, before pulling your hair off, you must keep in mind Linux is designed to be a multi-user environment. In an environment with more than one users, it is crucial to have a secure system for deciding which files are yours and who can fiddle with them.

Even if you're the only user on an ordinary desktop system, file permissions help keeping your important files safe, both from outsiders and your own mistakes.

2.2.2 Understanding File Ownership

Every file on your Linux system, including directories, is owned by a specific user and group. Therefore, file permissions are defined separately for users, groups, and others.

User: The username of the person who owns the file. By default, the user who creates the file will become its owner.

Group: The usergroup that owns the file. All users who belong into the group that owns the file will have the same access permissions to the file. This is useful if, for example, you have a project that requires a bunch of different users to be able to access certain files, while others can't. In that case, you'll add all the users into the same group, make sure the required files are owned by that group, and set the file's group permissions accordingly.

Other: A user who isn't the owner of the file and doesn't belong in the same group the file does. In other words, if you set a permission for the "other" category, it will affect everyone else by default. For this reason, people often talk about setting the "world" permission bit when they mean setting the permissions for "other."

2.2.3 Understanding File Permissions

There are three types of access permissions on Linux: read, write, and execute. These permissions are defined separately for the file's owner, group and all other users.

Read permission: On a regular file, the read permission bit means the file can be opened and read. On a directory, the read permission means you can list the contents of the directory.

Write permission: On a regular file, this means you can modify the file, aka write new data to the file. In the case of a directory, the write permission means you can add, remove, and rename files in the directory. This means that if a file has the write permission bit, you are allowed to modify the file's contents, but you're allowed to rename or delete the file only if the permissions of the file's *directory* allow you to do so.

Execute permission: In the case of a regular file, this means you can execute the file as a program or a shell script. On a directory, the execute permission (also called the "search bit") allows you to access files in the directory and enter it, with the cd command, for example. However, note that although the execute bit lets you enter the directory, you're not allowed to list its contents, unless you also have the read permissions to that directory.

2.2.4 How to view file permissions

You can view the access permissions of a file by doing the long directory listing with the ls -l command. This is what a long directory listing might look like:

```
me@puter:  /home/writers$ ls  -l
total  17
drwxr-xr-x  3  nana  writers  80  2005-09-20  21:37  dir
-rw-r--  1  nana  writers  8187  2005-09-19  13:35  file
-rwxr-xr-x  1  nana  writers  10348  2005-07-17  20:31  otherfile
```

What does the output of ls -l mean? The very first column, the one that looks like a bunch of mumbo jumbo, shows the file type and permissions. The second column shows the number of links (directory entries that refer to the file), the third one shows the owner of the file, and the fourth one shows the group the file belongs to. The other columns show the file's size in bytes, date and time of last modification, and the filename.

The first column, the one that shows the file's permissions and looks like mumbo jumbo, is organized into four separate groups, although it certainly doesn't look very organized.

The first group consists of only one character, and it shows the file's type. For example, d means a directory and - means a normal file, so if you take a look at our example output, you'll notice `dir` is a directory, while `file` and `otherfile` are regular files.

The first character can be any of these:

d = directory
- = regular file
l = symbolic link
s = Unix domain socket
p = named pipe
c = character device file
b = block device file

The next nine characters show the file's permissions, divided into three groups, each consisting of three characters. The first group of three characters shows the read, write, and execute permissions for *user*, the owner of the file. The next group shows the read, write, and execute permissions for the *group* of the file. Similarly, the last group of three characters shows the permissions for *other*, everyone else. In each group, the first character means the read permission, the second one write permission, and the third one execute permission.

The characters are pretty easy to remember.

r = read permission
w = write permission
x = execute permission
- = no permission

What does this mean in practice? Let's have an example. Remember the imaginary directory listing we did at the beginning? The output looked like this:

```
drwxr-xr-x 3   nana writers 80 2005-09-20  21:37  dir
-rw-r—  1      nana writers 8187  2005-09-19  13:35  file
-rwxr-xr-x 1   nana writers 10348 2005-07-17  20:31  otherfile
```

As we already noticed, dir is a directory, because the first column begins with a d. The owner of this directory is user nana and the group owner is writers. The first three characters, rwx, indicate the directory's owner, nana in this case, has full access to the directory. The user nana is able to access, view, and modify the files in that directory. The next three characters, r-x, indicate that all users belonging to group writers have read and execute permissions to the directory. They can change into the directory, execute files, and view its contents. However, because they don't have write permissions, they can't make any changes to the directory content. Finally, the last three characters, r-x, indicate that all the users who are not nana or don't belong into group writers, have read and execute permissions in the directory.

How about file? Because the first column begins with a -, the file is a regular file, owned by user nana and group writers, just like the directory in our example. The first three characters, rw-, indicate the owner has read and write access to the file. According to the next three characters, r—, the users belonging to group writers can view the file but not modify or execute it. The final three characters, —-, indicate no one else has any access to the file.

Similarly, you can see otherfile is a regular file and its owner has full access to it, while everyone else can read and execute the file but not modify it.

2.2.5 How to Set File Permissions—Symbolic Mode

You can set file permissions with the chmod command. Both the root user and the file's owner can set file permissions. chmod has two modes, symbolic and numeric.

The symbolic mode is pretty easy to remember. First, you decide if you set permissions for the user (u), the group (g), others (o), or all of the three (a). Then, you either add a permission (+), remove it (–), or wipe out the previous permissions and add a new one (=). Next, you decide if you set the read permission (r), write permission (w), or execute permission (x). Last, you'll tell chmod which file's permissions you want to change.

Let's have a couple of examples. Suppose we have a regular file called testfile, and the file has full access permissions for all the groups (long directory listing would show -rwxrwxrwx as the file's permissions).

Wipe out all the permissions but add read permission for everybody:

`$ chmod a=r testfile`

After the command, the file's permissions would be -r–r–r–

Add execute permissions for group:

`$ chmod g+x testfile`

Now, the file's permissions would be -r–r-xr–

Add both write and execute permissions for the file's owner. Note how you can set more than one permission at the same time:

`$ chmod u+wx testfile`

After this, the file permissions will be -rwxr-xr–

Remove the execute permission from both the file's owner and group. Note, again, how you can set them both at once:

`$ chmod ug-x testfile`

Now, the permissions are -rw-r–r–

As a summary, have a look at this quick reference for setting file permissions in symbolic mode:

Which user?	
u	user/owner
g	group
o	other
a	all

What to do?	
+	add this permission
–	remove this permission
=	set exactly this permission

Which permissions?	
r	read
w	write
x	execute

2.2.6 How to Set File Permissions - Numeric Mode

The other mode in which chmod can be used is the numeric mode. In the numeric mode, the file permissions aren't represented by characters. Instead, they are represented by a three-digit octal number.

4 = read (r)
2 = write (w)
1 = execute (x)
0 = no permission (-)

To get the permission bits you want, you add up the numbers accordingly. For example, the rwx permissions would be 4+2+1=7, rx would be 4+1=5, and rw would be 4+2=6. Because you set separate permissions for the owner, group, and others, you'll need a three-digit number representing the permissions of all these groups.

Let's have an example.

$ chmod 755 testfile

This would change the `testfile's` permissions to `-rwxr-xr-x`. The owner would have full read, write, and execute permissions (7=4+2+1), the group would have read and execute permissions (5=4+1), and the world would have the read and execute permissions as well.

Let's have another example:

$ chmod 640 testfile

In this case, `testfile's` permissions would be `-rw-r---`. The owner would have read and write permissions (6=4+2), the group would have read permissions only (4), and the others wouldn't have any access permissions (0).

The numeric mode may not be as straightforward as the symbolic mode, but with the numeric mode, you can more quickly and efficiently set the file permissions. This quick reference for setting file permissions in numeric mode might help:

Which number?	
0	—
1	—x
2	-w-
3	-wx
4	r—
5	r-x
6	rw-
7	rwx

2.3 PROCESSES

Process control and the ability for inter process communication is handled by the Linux kernel.

2.3.1 Tools For Working with Processes

- accton - Turns process accounting on and off. Uses the file /var/log/pacct. To turn it on type "accton /var/log/pacct". Use the command with no arguments to turn it off.
- kill - Kill a process by number
- killall - Send a signal to a process by name
- lastcomm (1) - Display information about previous commands in reverse order. Works only if process accounting is on.
- nice - Set process priority of new processes.
- ps(1) - Used to report the status of one or more processes.
- pstree(1) - Display the tree of running processes.
- renice(8) - Can be used to change the process priority of a currently running process.
- sa(8) - Generates a summary of information about users' processes that are stored in the /var/log/pacct file.
- skill - Report process status.
- snice - Report process status.

- top - Displays the processes that are using the most CPU resources.

2.3.2 Process Scheduling

Computer time on Linux systems is allocated in jiffies. A jiffie is a microprocessor time slice. On most Linux systems it is 1/100 of a second. On some systems it is 1/1024 of a second. The Linux kernel controls process scheduling. There are three types of scheduling:

1. normal - Referred to as other, this is the scheduling type set for normal programs.
2. FIFO - This is a real time scheduling priority. The FIFO term means the first process started (first in) will be the first done (first out). The only time this type of process exits is if it sleeps, is rescheduled, or if it must wait on other kernel priorities to be done.
3. RR - This is a round robin type of scheduling, where each task gets a certain amount of time, then it must exit, yield control to the next task and get back into the task queue. This is a real time scheduling priority.

Linux processes have the following characteristics:

1. policy - normal or real time. Real time processes have a higher priority than normal processes.
2. priority - The process priority. It is a number between -20 and 19. The value of -20 is the highest, and 19 is the lowest priority. Process priority can be set with the nice(1) command and changed using the renice(8) command.

2.3.3 Inter-Process Communication

The types of inter process communication are:

1. **Signals:** Sent by other processes or the kernel to a specific process to indicate various conditions.
2. **Pipes:** Unnamed pipes set up by the shell normally with the "|" character to route output from one program to the input of another.
3. **FIFOS:** Named pipes operating on the basis of first data in, first data out.
4. **Message queues:** Message queues are a mechanism set up to allow one or more processes to write messages that can be read by one or more other processes.
5. **Semaphores:** Counters that are used to control access to shared resources. These counters are used as a locking mechanism to prevent more than one process from using the resource at a time.
6. **Shared memory:** The mapping of a memory area to be shared by multiple processes.

Message queues, semaphores, and shared memory can be accessed by the processes if they have access permission to the resource as set up by the object's creator. The process must pass an identifier to the kernel to be able to get the access.

2.3.4 Signals

Linux Signals are:

Signal Name	Number	Description
SIGHUP	1	Hangup (POSIX)
SIGINT	2	Terminal interrupt (ANSI)
SIGQUIT	3	Terminal quit (POSIX)
SIGILL	4	Illegal instruction (ANSI)

Signal Name	Number	Description
SIGTRAP	5	Trace trap (POSIX)
SIGIOT	6	IOT Trap (4.2 BSD)
SIGBUS	7	BUS error (4.2 BSD)
SIGFPE	8	Floating point exception (ANSI)
SIGKILL	9	Kill(can't be caught or ignored) (POSIX)
SIGUSR1	10	User defined signal 1 (POSIX)
SIGSEGV	11	Invalid memory segment access (ANSI)
SIGUSR2	12	User defined signal 2 (POSIX)
SIGPIPE	13	Write on a pipe with no reader, Broken pipe (POSIX)
SIGALRM	14	Alarm clock (POSIX)
SIGTERM	15	Termination (ANSI)
SIGSTKFLT	16	Stack fault
SIGCHLD	17	Child process has stopped or exited, changed (POSIX)
SIGCONT	18	Continue executing, if stopped (POSIX)
SIGSTOP	19	Stop executing(can't be caught or ignored) (POSIX)
SIGTSTP	20	Terminal stop signal (POSIX)
SIGTTIN	21	Background process trying to read, from TTY (POSIX)
SIGTTOU	22	Background process trying to write, to TTY (POSIX)
SIGURG	23	Urgent condition on socket (4.2 BSD)
SIGXCPU	24	CPU limit exceeded (4.2 BSD)
SIGXFSZ	25	File size limit exceeded (4.2 BSD)
SIGVTALRM	26	Virtual alarm clock (4.2 BSD)
SIGPROF	27	Profiling alarm clock (4.2 BSD)
SIGWINCH	28	Window size change (4.3 BSD, Sun)
SIGIO	29	I/O now possible (4.2 BSD)
SIGPWR	30	Power failure restart (System V)

As noted above, processes can ignore, block, or catch all signals except SIGSTOP and SIGKILL. If a process catches a signal, it means that it includes code that will take appropriate action when the signal is received. If the signal is not caught by the process, the kernel will take default action for the signal.

2.3.5 FIFOs

FIFOs are permanent objects and can be created using the mkfifo(1) or mknod(1) command. Inside the program, the FIFO can be created using the mknod command, then opened and read from or written to just like a normal file. The FIFO is normally in blocking mode when attempting to perform read operations.

2.3.6 Multi-user and Multi-Tasking

Now that we are more used to our environment and we are able to communicate a little bit with our system, it is time to study the processes we can start in more detail. Not every command starts a single process. Some commands initiate a series of processes, such as **mozilla**; others, like ls, are executed as a single command.

Furthermore, Linux is based on UNIX, where it has been common policy to have multiple users running multiple commands, at the same time and on the same system. It is obvious that measures have to be taken to have the CPU manage all these processes, and that functionality has to be provided so users can switch between processes. In some cases, processes will have to continue to run even when the user who started them logs out. And users need a means to reactivate interrupted processes.

2.3.7 Interactive Processes

Interactive processes are initialized and controlled through a terminal session. In other words, there has to be someone connected to the system to start these processes; they are not started automatically as part of the system functions. These processes can run in the foreground, occupying the terminal that started the program, and you can't start other applications as long as this process is running in the foreground. Alternatively, they can run in the background, so that the terminal in which you started the program can accept new commands while the program is running. Until now, we mainly focussed on programs running in the foreground – the length of time taken to run them was too short to notice - but viewing a file with the less command is a good example of a command occupying the terminal session. In this case, the activated program is waiting for you to do something. The program is still connected to the terminal from where it was started, and the terminal is only useful for entering commands this program can understand. Other commands will just result in errors or

While a process runs in the background, however, the user is not prevented from doing other things in the terminal in which he started the program, while it is running.

The shell offers a feature called *job control* which allows easy handling of multiple processes. This mechanism switches processes between the foreground and the background. Using this system, programs can also, be started in the background immediately.

Running a process in the background is only useful for programs that don't need user input (via the shell), Putting a job in the background is typically done when execution of a job is expected to take a long time. In order to free the issuing terminal after entering the command, a trailing ampersand is added. In the example, using graphical mode, we open an extra terminal window from the existing one:

```
billy:->    xterm 6
[1] 26558

billy:->jobs
[1]+ Running   xterm 6
```

The full job control features are explained in detail in the hash Info pages, so only the frequently used job control applications arc listed here:

(part of) command	Meaning
regular_command	Runs this command in the foreground.
command &	Run this command in the background (release the terminal)
jobs	Show commands running in the background.
Ctrl+Z	Suspend (stop, but not quit) a process running in the foreground (suspend).

Clrl+C	Interrupt (terminate and quit) a process running in the foreground.
%n	Every process running in the background gets a number assigned to it. By using the % expression a job can be referred to using its number. for instance fg *%2.*
bg	Reactivate a suspended program in the background.
fg	Puts the job back in the foreground.
kill	End a process (also see Shell Builtin Commands in the Info pages of hash)

More practical examples can be found in the exercises.

Most UNIX systems are likely to be able to run screen, which is useful when you actually want another shell to execute commands. Upon calling screen, a new session is created with an accompanying shell and/or commands as specified, which you can then put out of the way. In this new session you may do whatever it is you want to do. All programs and operations will run independent of the issuing shell. You can then detach this session, while the programs you started in it continue to run, even when you log out of the originating shell, and pick your *screen* up again any time you like.

This program originates from a time when virtual consoles were not invented yet, and everything needed to be done using one text terminal. To addicts, it still has meaning in Linux, even though we've had virtual consoles for almost ten years.

2.3.8 Process Attributes

A process has a series of characteristics, which can be viewed with the ps command:

- The process ID or PID: a unique identification number used to refer to the process.
- The parent process ID or PPID: the number of the process (PID) that started this process.
- Nice number the degree of friendliness of this process toward other processes (not to be confused with process priority, which is calculated based on this nice number and recent CPU usage of the process).
- Terminal or TTY: terminal to which the process is connected.
- User name of the real and effective user (RUID and EUID): the owner of the process. The real owner is the user issuing the command, the effective user is the one determining access to system resources. RUID and EUID are usually the same, and the process has the same access rights the issuing user would have. An example to clarify this: the browser mozilla in /usr/bin is owned by user *root:*

```
theo:->   Is -1 /usr/bin/mozilla
-rwxr-xr-x 1 root root          4996 Nov 20 18:28 /usr/bin/mozilla*
theo: > mozilla 6
[1] 26595
theo:-> ps -af
UID   PID  PPID C STIKE TTY  TIME     CMD
theo 26601 26599 0 15:04 pts/5 00:00:00 /usr/lib/mozilla/mozilla-bin
theo 26613 26569 0 15:04 pts/S 00:00:00 ps -af
```

When user *theo* starts this program, the process itself and all processes started by the initial process, will be owned by user *theo* and not by the system administrator. When mozilia needs access to certain files, that access will be determined by *theo's.* permissions and not by *roots.*

- Real and effective group owner (RGID and EGID): The real group owner of a process is the primary group of the user who started the process. The effective group owner is usually the same, except when SGID access mode has been applied to a file.

2.4 PROCESS STATUS

The ps command is one of the tools for visualizing processes. This command has several options which can be combined to display different process attributes.

With no options specified, ps only gives information about the current shell and eventual processes:

```
theo:-> ps
PID    TTY      TIME      CMD
4245   pcs/7    00:00:00  bash
5314   pC3/7    00:00:00  ps
```

Since, this does not give enough information - generally, at least a hundred processes are running on your system - we will usually select particular processes out of the list of all processes, using the grep command in *a pipe.* See Section 5 1.11. as in this line, which will select and display all processes owned by a particular user.

ps -ef | grep *username*

This example shows all processes with a process name of bash, the most common login shell on Linux systems:

```
theo:> ps  auxw    | grap  bash
brenda      31970  0.0  0.3  6030  1556  tty2  S  Feb23  0:00   -bash
root        32043  0.0  0.3  6112  1600  tty4  S  Feb23  0:00   -bash
theo        32581  0.0  0.3  6384  1864  pts/1 s  Feb23  0:00   bash
theo        32616  0.0  0.3  6396  1896  pts/2 s  Feb23  0:00   bash
theo        32629  0.0  0.3  380   1856  pts/3  s Feb23  0:00   bash
theo        2214   0.0  0.3  6412  1944  pts/5  s 16:18  0:02   bash
theo        4245   0.0  0.3  6392  1888  pts/7  s 17:26  0:00   bash
theo        5427   0.0  0.1  3720  548   pts/7   s  19:22 0:00  grep bash
```

In these cases, the grep command finding lines containing the string *bash* is often displayed as well on systems that have a lot of idletime. If you don't want this to happen, use the pgrep command.

Bash shells are a special case: this process list also shows which ones are login shells (where you have to give your usename and password, such as when you log in in textmode or do a remote login, as opposed to non-login shells, started up for instance by clicking a terminal window icon). Such, login shells are preceded with a dash (-).

2.4.1 Life and death of a process

2.4.1.1 Process creation

A new process is created because an existing process makes an exact copy of itself. This child process has the same environment as its parent, only the process ID number is different. This procedure is called *forking.*

After the forking process, the address space of the child process is overwritten with the new process data. This is done through an *exec* call to the system.

The *fork-and-exec* mechanism thus switches an old command with a new, while the environment in which the new program is executed remains the same, including configuration of input and output devices, environment variables and priority. This mechanism is used to create all UNIX processes, so it also applies to the Linux operating system. Even the first process, init, with process ID 1, is forked during the boot procedure in the so-called *bootstrapping* procedure.

This scheme illustrates the fork-and-exec mechanism. The process ID changes after the fork procedure:

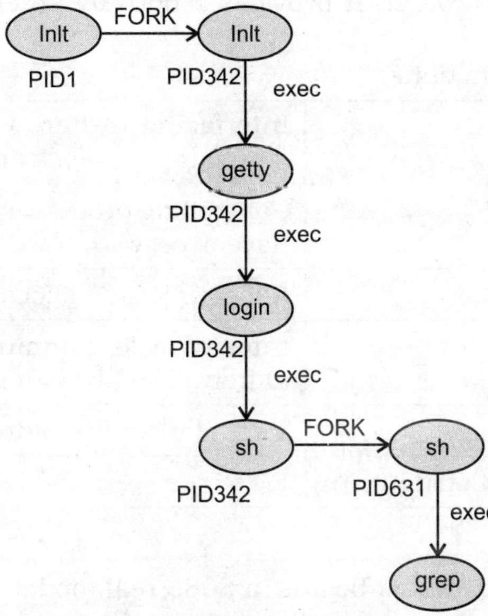

There are a couple of cases in which init becomes the parent of a process, while the process was not started by init, as we already saw in the pstree example. Many programs, for instance, *daemonize* their child processes, so they can keep on running when the parent stops or is being stopped. A window manager is a typical example: it starts an xterm process that generates a shell that accepts commands. The window manager then denies any further responsibility and passes the child process to init. Using this mechanism, it is possible to change window managers without interrupting running applications.

Every now and then things go wrong, even in good families. In an exceptional case, a process might finish while the parent does not wait for the completion of this process. Such an unburied process is called a *zombie* process.

2.5 KILLING PROCESS

Processes can be stopped by killing them. kill -9 <Process ID> kills process.

To be successful use 9 the hard way. To find out that you are probably to nice use 150 the soft way.KDE system guard is an application to show and kill processes.

2.6 PROCESS PRIORITY

2.7 LINUX KERNEL

The best document I have ever found to explain the Linux kernel in everyday readable language is "The Linux Kernel" by David A Rusling. Another excellent source is "A Tour of the Linux Kernel Source" by Alessandro Rubini. If you want more information, read the Linux kernel source code. This section gives a brief synopsis of some functions performed by the Linux kernel in an effort to help the user use the system through better understanding.

The kernel acts as a mediator for your programs and your hardware. First, it performs memory management for all of the running programs, and manages the time slices of the processor's cycles that they get. It provides a portable interface for programs to talk to your hardware.

The kernels main functions:

Device drivers:	Interfacing to hardware through device drivers for character, block and network interface devices
Process management:	Controlling processes and the address space they have access to
	Allocating time slices for processes
	Inter-process communication including process to network card communication
Memory management:	Virtual memory addressing control
Filesystem control and structuring	
Networking	

When the kernel is loaded, it begins in 8086 real mode. It moves parts of itself to two different addresses. The kernel identifies some of the hardware characteristics of the system. At this point, it may ask the user to choose the video mode they want to run the console at. Then it moves its system area from an address higher up in memory to 1000 hexadecimal. Then it enters protected mode and decompresses itself. It stores the decompressed code and data, begins execution of the decompressed code, sets up the processors register tables for memory management, and sets up memory paging.

The following parts of the kernel are initialized:

- Memory bounds are set
- The traps, IRQ channels and scheduling are initialized
- The command line is parsed
- The device drivers and disk buffering are initialized
- The delay loop, called the BogoMips number is calculated
- Tests the to see if interrupt 16 works with the coprocessor

The kernel begins user mode and forks the init process.

The init process tries to run the first of the following programs it can find:

`/sbin/init, /etc/init, bin/init, or /bin/sh.`

If the last is run, it forks a root shell on the first terminal.

2.7.1 Kernel Module Support

Most kernels (except those for floppy boot disks or small remote systems) are compiled so modular support is required.

The package modules.tar.gz contains all the programs needed to manage modules. This should already be installed on most distributions. The kernel modules are usually in a directory pertinent to the kernel version in /lib/modules. Modules can be found in "lib/modules/2.2.12-20" for kernel version 2.2.12-20. They are loadable modules ending in ".o" that are used to support the kernel.

To load a module type "insmod module" where "module" is the name of the module to load. Ex: insmod /lib/modules/2.2.12-20/misc/ftape.o

Programs used to manage modules are:

- insmod - Installs a loadable kernel module into the running kernel.
- lsmod - Lists all the currently loaded kernel modules
- rmmod - Unloads modules, Ex: rmmod ftape
- depmod - Creates a dependency file, "modules.dep" in the directory "/lib/modules/x.x.x", later used by modprobe to automatically load the relevant modules.
- modprobe - Used to load a module or set of modules. Loads all modules specified in the file "modules.dep".

Modules are loaded from startup script files using "modprobe" to handle loadable modules automatically.

FILES:

- /etc/conf.modules - A list of alias names for modules used to help determine system required modules. See the man page on depmod.

modprobe -l |more	Lists all the modules available for your kernel
rmmod module_name	Remove a module from the kernel

2.7.2 Kernel Parameter Modification

The "sysctl" program is a tool used to modify kernel parameters, or what is actually variables and data structures. If you type "sysctl -a |more" you will see a long list of kernel parameters. You can use this sysctl program to modify these parameters. However, I have been unable to add new parameters.

2.7.3 Starting the Kernel

In order for the kernel to be properly started, it must be in the correct location as pointed to by the lilo boot loader program. This location is specified in the /etc/lilo.conf configuration as described above. Also the kernel must be properly setup after any new compile is done. The compile is done from the directory /usr/src/linux. The boot image for the new kernel should be taken from the file "/usr/src/linux/arch/i386/boot/bzImage" and renamed vmlinuz or vmlinux-2.2.12-20 or whatever it is your intention to call it. Sometimes a softlink file in /boot called vmlinuz points to the kernel that is to be run. If the copy of vmlinuz is copied from the directory /usr/src/linux, the kernel may not load properly. Sometimes the error message states that the kernel is too large.

2.8 STARTING AND SHUT DOWN PERIPHERALS

2.8.1 What Are Standard Input and Standard Output?

Most Linux commands read input, such as a file or another attribute for the command, and write output. By default, input is being given with the keyboard, and output is displayed on your screen. Your keyboard is your *standard input* (stdin) device, and the screen or a particular terminal window is the *standard output* (stdout) device.

However, since Linux is a flexible system, these default settings don't necessarily have to be applied. The standard output, for example, on a heavily monitored server in a large environment may be a printer.

Sometimes you will want to put output of a command in a file, or you may want to issue another command on the output of one command. This is known as redirecting output. Redirection is done using either the ">" (greater-than symbol), or using the "l" (pipe) operator which sends the standard output of one command to another command as standard input.

As we saw before, the cat command concatenates files and puts them all together to the standard output. By redirecting this output to a file, this file name will be created -or overwritten if it already exists, so take care,

```
nancy:-> cat test1
some words
nancy:-> cat test2
some other words
nancy:-> cat test1 test2 > test3
nancy:-> cat test3
some words
```

Redirecting "nothing" to an existing file is equal to emplying the file:

```
nancy:-> 1s -1list
-rw-rw-r-- 1 nancy nancy 117 Apr 2 18:09 list
nancy:-> list
nancy:-> 1s -1 list
-rw-rw-r-- 1 nincy nancy 0 Apr 412:01 list
```

This process is called *tmcamit.*

The same redirection to an nonexistent file will create a new empty file with the given name:

```
nancy:-> 1s -1 newlist
1st: newlist: No such file or directory
nancy:-> > newlist
nancy:-> 1s -1 newlist
-rw-rw-r– 1 nancy nancy 0 Apr 4 12:05 newlist
```

Some examples using piping of commands:

To find a word within some text, display all lines matching "pattern1", and exclude lines also matching "pattern2" from being displayed:

```
greppatternl  file|grep-vpattern2
```

To display output of a directory listing one page at a time:

```
Is-la  |less
```

To find a file in a directory:

```
Is-lgrep  part_of_file_name
```

2.9 BIOS

The BIOS is located in a memory chip of the motherboard and is the program that is started when the PC boots. It performs checks and signals failures with beeps but also puts error codes to the IO port address 80 that can be captured using special PC cards. After a successful boot, it looks for a disk drive as a hard disk and reads there the boot sector and starts it. Usually a boot loader as Grub has been installed there and comes alive.

To get a nice boot sequence seeing one tux per processor, you need to enable framebuffer support in the kernel, select therefore VESA VGA graphics and Framebuffer Console support.

2.10 CONSOLE

The **Linux console** is a **system console** support in the **Linux kernel**. The Linux console was the first functionality of the kernel, developed as early as in 1991 (see **history of Linux**). On **PC** architecture, it is common to use **VGA**-compatible video hardware. Implementations of **computer graphics** on Linux are excluded from the scope of this article. Linux console, like pure text mode, uses **monospace fonts**.

Virtual consoles allow the storage of multiple text buffers, enabling different console programs to run simultaneously but interact with the user in different contexts. From the user's point of view, this creates the illusion of several independent consoles.

Each virtual console can have its own character set and keyboard layout. Linux 2.6 introduced the ability to load a different font for each virtual console (kernel versions predating 2.6 change the font only on demand).

2.11 THE SCHEDULER

2.11.1 Linux Scheduler

The job of a scheduler is to arbitrate access to the current CPU between multiple processes. The scheduler is implemented in the 'main kernel file' kernel/sched.c. The corresponding header file include/linux/sched.h is included (either explicitly or indirectly) by virtually every kernel source file.

The fields of task structure relevant to scheduler include:

- p->need_resched: this field is set if schedule() should be invoked at the 'next opportunity'.
- p->counter: number of clock ticks left to run in this scheduling slice, decremented by a timer. When this field becomes lower than or equal to zero, it is reset to 0 and p->need_resched is set. This is also sometimes called 'dynamic priority' of a process because it can change by itself.
- p->priority: the process' static priority, only changed through well-known system calls like **nice(2)**, POSIX.1b **sched_setparam(2)** or 4.4BSD/SVR4 **setpriority(2)**.

- p->rt_priority: realtime priority
- p->policy: the scheduling policy, specifies which scheduling class the task belongs to. Tasks can change their scheduling class using the **sched_setscheduler(2)** system call. The valid values are SCHED_OTHER (traditional UNIX process), SCHED_FIFO (POSIX.1b FIFO realtime process) and SCHED_RR (POSIX round-robin realtime process). One can also OR SCHED_YIELD to any of these values to signify that the process decided to yield the CPU, for example by calling **sched_yield(2)** system call. A FIFO realtime process will run until either a) it blocks on I/O, b) it explicitly yields the CPU or c) it is preempted by another realtime process with a higher p->rt_priority value. SCHED_RR is the same as SCHED_FIFO, except that when its timeslice expires it goes back to the end of the runqueue.

The scheduler's algorithm is simple, despite the great apparent complexity of the schedule() function. The function is complex because it implements three scheduling algorithms in one and also because of the subtle SMP-specifics.

The apparently 'useless' gotos in schedule() are there for a purpose - to generate the best optimized (for i386) code. Also, note that scheduler (like most of the kernel) was completely rewritten for 2.4, therefore the discussion below does not apply to 2.2 or earlier kernels.

Let us look at the function in detail:

1. If current->active_mm == NULL then something is wrong. Current process, even a kernel thread (current->mm == NULL) must have a valid p->active_mm at all times.
2. If there is something to do on the tq_scheduler task queue, process it now. Task queues provide a kernel mechanism to schedule execution of functions at a later time. We shall look at it in details elsewhere.
3. Initialize local variables prev and this_cpu to current task and current CPU respectively.
4. Check if schedule() was invoked from interrupt handler (due to a bug) and panic if so.
5. Release the global kernel lock.
6. If there is some work to do via softirq mechanism, do it now.
7. Initialize local pointer struct schedule_data *sched_data to point to per-CPU (cacheline-aligned to prevent cacheline ping-pong) scheduling data area, which contains the TSC value of last_schedule and the pointer to last scheduled task structure (TODO: sched_data is used on SMP only but why does init_idle() initializes it on UP as well?).
8. runqueue_lock spinlock is taken. Note that we use spin_lock_irq() because in schedule() we guarantee that interrupts are enabled. Therefore, when we unlock runqueue_lock, we can just re-enable them instead of saving/restoring eflags (spin_lock_irqsave/restore variant).
9. task state machine: if the task is in TASK_RUNNING state, it is left alone; if it is in TASK_INTERRUPTIBLE state and a signal is pending, it is moved into TASK_RUNNING state. In all other cases, it is deleted from the runqueue.
10. next (best candidate to be scheduled) is set to the idle task of this cpu. However, the goodness of this candidate is set to a very low value (-1000), in hope that there is someone better than that.
11. If the prev (current) task is in TASK_RUNNING state, then the current goodness is set to its goodness and it is marked as a better candidate to be scheduled than the idle task.
12. Now the runqueue is examined and a goodness of each process that can be scheduled on this cpu is compared with current value; the process with highest goodness wins.

Now the concept of "can be scheduled on this cpu" must be clarified: on UP, every process on the runqueue is eligible to be scheduled; on SMP, only process not already running on another cpu is eligible to be scheduled on this cpu. The goodness is calculated by a function called goodness(), which treats realtime processes by making their goodness very high (1000 + p->rt_priority), this being greater than 1000 guarantees that no SCHED_OTHER process can win; so they only contend with other realtime processes that may have a greater p->rt_priority. The goodness function returns 0 if the process' time slice (p->counter) is over. For non-realtime processes, the initial value of goodness is set to p->counter - this way, the process is less likely to get CPU if it already had it for a while, i.e. interactive processes are favoured more than CPU bound number crunchers. The arch-specific constant PROC_CHANGE_PENALTY attempts to implement "cpu affinity" (i.e. give advantage to a process on the same CPU). It also gives a slight advantage to processes with mm pointing to current active_mm or to processes with no (user) address space, i.e. kernel threads.

13. if the current value of goodness is 0 then the entire list of processes (not just the ones on the runqueue!) is examined and their dynamic priorities are recalculated using simple algorithm:

```
recalculate:
        {
            struct  task_struct  *p;
            spin_unlock_irq(&runqueue_lock);
            read_lock(&tasklist_lock);
            for_each_task(p)
            p->counter  =  (p->counter  >>  1)  +  p->priority;
            read_unlock(&tasklist_lock);
            spin_lock_irq(&runqueue_lock);
        }
```

Note that the we drop the runqueue_lock before we recalculate. The reason is that we go through entire set of processes; this can take a long time, during which the schedule() could be called on another CPU and select a process with goodness good enough for that CPU, whilst we on this CPU were forced to recalculate. Ok, admittedly this is somewhat inconsistent because while we (on this CPU) are selecting a process with the best goodness, schedule() running on another CPU could be recalculating dynamic priorities.

14. From this point on it is certain that next points to the task to be scheduled, so we initialize next->has_cpu to 1 and next->processor to this_cpu. The runqueue_lock can now be unlocked.
15. If we are switching back to the same task (next == prev), then we can simply reacquire the global kernel lock and return, i.e. skip all the hardware-level (registers, stack, etc.) and VM-related (switch page directory, recalculate active_mm etc.) stuff.
16. The macro switch_to() is architecture specific. On i386, it is concerned with (a) FPU handling, (b) LDT handling, (c) reloading segment registers, (d) TSS handling, and (e) reloading debug registers.

2.12 THE INIT PROGRAM

As seen in the previous section, the kernel will start a program called init, if it finds it.

The init process reads the file "/etc./inittab" and uses this file to determine how to create processes. Read the init man page for more information. Also note that init is always running and can dynamically do things and run processes based upon various signals. The administrator can also cause it to dynamically change system processes and runlevels by using the telinit program or editing the "/etc/inittab" file.

2.13 RUNLEVELS

Linux utilizes what is called "runlevels". A runlevel is a software configuration of the system that allows only a selected group of processes to exist. Init can run the system in one of eight runlevels. These runlevels are 0-6 and S or s. The system runs in only one of these runlevels at a time. Typically these runlevels are used for different purposes. Runlevels 0, 1, and 6 are reserved. For Redhat Linux version 6, the runlevels are:

0	–	halt
1	–	Single user mode
2	–	Multiuser, without NFS (The same as 3, if you don't have networking)
3	–	Full multiuser mode
4	–	unused
5	–	X11
6	–	Reboot

2.14 THE INITTAB FILE

The "/etc/inittab" file tells init which runlevel to start the system at and describes the processes to be run at each runlevel. An entry in the inittab file has the following format:

id:runlevels:action:process

- id - A unique sequence of 1-4 characters which identifies an entry in inittab.
- runlevels - Lists the runlevels for which the specified action should be taken. This field may contain multiple characters for different runlevels allowing a particular process to run at multiple runlevels. For example, 123 specifies that the process should be started in runlevels 1, 2, and 3.
- action - Describes which action should be taken. Valid actions are listed below
  - respawn - The process will be restarted whenever it terminates.
  - wait - The process will be started once when the specified runlevel is entered and init will wait for its termination.
  - once - The process will be executed once when the specified runlevel is entered
  - boot - The process will be executed during system boot. The runlevels field is ignored.
  - bootwait - Same as "boot" above, but init waits for its termination.
  - off - This does nothing.
  - ondemand - This process will be executed whenever the specified ondemand runlevel is called.
  - initdefault - Specifies the runlevel which should be entered after system boot. If none exists, init will ask for a runlevel on the console. The process field is ignored.
  - sysinit - The process will be executed during system boot. It will be executed before any boot or bootwait entries. The runlevels field is ignored.

- powerwait - The process will be executed when init receives the SIGPWR signal. Init will wait for the process to finish before continuing.
- powerfail - Same as powerwait but init does not wait for the process to complete.
- powerokwait - The process will be executed when init receives the SIGPWR signal provided there is a file called "/etc/powerstatus" containing the word "OK". This means that the power has come back again.
- ctrlaltdel - This process is executed when init receives the SIGINT signal. This means someone on the system console has pressed the "CTRL-ALT-DEL" key combination.
- kbrequest - The process will be executed when init receives a signal from the keyboard handler that a special key combination was pressed on the console keyboard.
- process - Specifies the process to be executed. If the process starts with the '+' character, init will not do utmp and wtmp accounting for that process. This is needed for gettys that insist on doing their own utmp/wtmp housekeeping (a historic bug).

Below is an example file:

```
# inittab This file describes how the INIT process should set up
# the system in a certain run-level.
#
# Author: Miquel van Smoorenburg, <miquels@drinkel.nl.mugnet.org>
# Modified for RHS Linux by Marc Ewing and Donnie Barnes
#

# Default runlevel. The runlevels used by RHS are:
# 0 - halt (Do NOT set initdefault to this)
# 1 - Single user mode
# 2 - Multiuser, without NFS (The same as 3, if you do not have
networking)
# 3 - Full multiuser mode
# 4 - unused
# 5 - X11
# 6 - reboot (Do NOT set initdefault to this)
#
1) id:3:initdefault:
# System initialization.
2) si::sysinit:/etc/rc.d/rc.sysinit
3) l0:0:wait:/etc/rc.d/rc 0
4) l1:1:wait:/etc/rc.d/rc 1
5) l2:2:wait:/etc/rc.d/rc 2
6) l3:3:wait:/etc/rc.d/rc 3
7) l4:4:wait:/etc/rc.d/rc 4
8) l5:5:wait:/etc/rc.d/rc 5
9) l6:6:wait:/etc/rc.d/rc 6
# Things to run in every runlevel.
10)ud::once:/sbin/update
```

```
# Trap CTRL-ALT-DELETE
11)ca::ctrlaltdel:/sbin/shutdown -t3 -r now
# When our UPS tells us power has failed, assume we have a few minutes
# of power left. Schedule a shutdown for 2 minutes from now.
# This does, of course, assume you have powerd installed and your
# UPS connected and working correctly.
12)pf::powerfail:/sbin/shutdown -f -h +2 "Power Failure; System Shutting
Down"
# If power was restored before the shutdown kicked in, cancel it.
13)pr:12345:powerokwait:/sbin/shutdown -c "Power Restored; Shutdown
Cancelled"
# Run gettys in standard runlevels
14)1:2345:respawn:/sbin/mingetty tty1
15)2:2345:respawn:/sbin/mingetty tty2
16)3:2345:respawn:/sbin/mingetty tty3
17)4:2345:respawn:/sbin/mingetty tty4
18)5:2345:respawn:/sbin/mingetty tty5
19)6:2345:respawn:/sbin/mingetty tty6
# Run xdm in runlevel 5
# xdm is now a separate service
20)x:5:respawn:/etc/X11/prefdm -nodaemon
```

On the left side of the file listing, above, are added numbers to help describe lines. Those lines without line numbers are either blank or begin with a "#" which means the line is a comment. Those line numbers are not part of the original file and are added here for reference purposes.

- On line 1 above you see "id:3:initdefault:". The id is "id" which stands for initdefault. Note that it is unique on all the numbered lines. The runlevel is 3 which sets the default starting runlevel to runlevel 3. The action is initdefault which tells init to make this runlevel the default runlevel. Note that the process field is blank since it is ignored by the initdefault action.
- Line 2 tells init to run the program "/etc/rc.d/rc.sysinit" during system boot, before any other processes.
- Lines 3 through 9 tell init to run the program "/etc/rc.d/rc" for runlevels 0 through 6. Note that for each line the appropriate runlevel is passed to the "/etc/rc.d/rc" script program on the command line. For example note on line 5 above the second field is the runlevel specifying 2. At the end of the line there is a space and a 2 which allows the variable 2 to be passed on the command line to the program.
- Line 10 specifies that the program "/sbin/update" will run once for every runlevel
- Line 11 sets up the program "/sbin/shutdown" to run when someone on the system console has pressed the "CTRL-ALT-DEL" key combination.
- Line 12 specifies "/sbin/shutdown" to run if the power fails. Note that there are different options passed on the command line for lines 11 and 12 although they run the same program.
- Line 13 specified "/sbin/shutdown" will run if power is restored for any of runlevels 1 through 5.
- Lines 14 through 19 specifies the "/sbin/mingetty" program to run on 6 different terminals for runlevels 2 through 5. This means that you can run 6 virtual terminals

from your keyboard simultaneously by pressing "ALT-F1" through "ALT-F6". Note pressing "ALT-F7" or above will do nothing, but the screen will not change from your current terminal.

Note the order of programs to run as specified above are:

1. /etc/rc.d/rc.sysinit
2. /etc/sbin/update
3. /etc/rc.d/rc 3 - Note: we are running runlevel 3 here.

2.15 RUN-LEVELS AND RUN LEVEL SCRIPTS

The runlevel scripts are used to bring up many system and networking functions. Since some functions are interdependent on other functions there is some required order in which these scripts must be run in order to bring the system up and to bring it gracefully down. Each runlevel has its own set of start(S) and kill(K) scripts but all these scripts are supported in the directory /etc/rc.d/init.d. This is because the start and kill scripts are soft links to the files in the /etc/rc.d/init.d directory.

The rc script Program

The script file /etc/rc.d/rc is run for the appropriate runlevel (typically 3 or 5) This file does the following:

1. It gets the previous and current system runlevels.
2. If the word confirm is in the file "/proc/cmdline" if sets up to run the scripts below in user confirmation mode.
3. All kill files (files whose first letter is 'K') in the subdirectory "/etc/rc.d/rc3.d" (assuming the previous runlevel was 3) are run. The parameter stop is usually passed on the command line to the kill script.
4. All startup files (files whose first letter is 'S") in the subdirectory "/etc/rc.d/rc5.d" (assuming the current runlevel is 5) are run. The parameter start is usually passed on the command line to the kill script.

These runlevel scripts are used to bring up (or down) various system services such as cron and gpm along with networking services from the network cards through Samba, and servers like DNS, DHCP, and NFS. A directory listing of the files in the /etc/rc.d/init.d will reveal the many possible services that the system can support.

```
drwxr-xr-x 2 root root 4096 May 5 10:00 .
drwxr-xr-x 10 root root 4096 Apr 28 04:08 ..
-rwxr-xr-x 1 root root 766 Sep 13 1999 amd
-rwxr-xr-x 1 root root 1231 Sep 20 1999 apmd
-rwxr-xr-x 1 root root 827 Sep 9 1999 arpwatch
-rwxr-xr-x 1 root root 989 Aug 16 1999 atd
-rwxr-xr-x 1 root root 4816 Sep 20 1999 autofs
-rwxr-xr-x 1 root root 1011 Dec 20 08:21 bootparamd
-rw--- 1 root root 1101824 Mar 2 11:25 core
-rwxr-xr-x 1 root root 1031 Sep 10 1999 crond
-rwxr-xr-x 1 root root 975 Apr 28 04:19 dhcpd
-rwxr-xr-x 1 root root 7386 Sep 20 1999 functions
-rwxr-xr-x 1 root root 1417 Aug 16 1999 gated
-rwxr-xr-x 1 root root 1261 Sep 24 1999 gpm
```

```
-rwxr-xr-x 1 root root 3129 Sep 20 1999 halt
-rwxr-xr-x 1 root root 865 Sep 21 1999 httpd
-rwxr-xr-x 1 root root 1151 Sep 13 1999 identd
-rwxr-xr-x 1 root root 1463 Sep 10 1999 inet
-rwxr-xr-x 1 root root 1924 Aug 30 1999 innd
-rwxr-xr-x 1 root root 6029 Sep 24 1999 isdn
-rwxr-xr-x 1 root root 1203 Sep 5 1999 keytable
-rwxr-xr-x 1 root root 449 Sep 11 1999 killall
-rwxr-xr-x 1 root root 1172 Sep 24 1999 kudzu
-rwxr-xr-x 1 root root 1890 Sep 13 1999 ldap
lrwxrwxrwx 1 root root 43 Dec 17 05:25 linuxconf -> /usr/lib/linuxconf/
redhat/scripts/linuxconf
-rwxr-xr-x 1 root root 1176 Sep 10 1999 lpd
-rwxr-xr-x 1 root root 1104 Sep 10 1999 mars-nwe
-rwxr-xr-x 1 root root 1171 Sep 24 1999 mcserv
-rwxr-xr-x 1 root root 1331 Sep 24 1999 named
-rwxr-xr-x 1 root root 3217 Sep 20 1999 netfs
-rwxr-xr-x 1 root root 6573 Sep 21 1999 network
-rwxr-xr-x 1 root root 2257 Sep 24 1999 nfs
-rwxr-xr-x 1 root root 1722 Sep 24 1999 nfslock
-rwxr-xr-x 1 root root 1603 Sep 20 1999 nscd
-r-xr-xr-x 1 root root 3439 Sep 27 1999 pcmcia
-rwxr-xr-x 1 root root 1086 Sep 10 1999 portmap
-rwxr-xr-x 1 root root 2435 Sep 26 1999 postgresql
-rwxr-xr-x 1 root root 1260 Sep 25 1999 pulse
-rwxr-xr-x 1 root root 955 Sep 26 1999 pxe
-rwxr-xr-x 1 root root 1532 Feb 4 1999 random
-rwxr-xr-x 1 root root 1270 Sep 10 1999 routed
-rwxr-xr-x 1 root root 780 Sep 22 1999 rstatd
-rwxr-xr-x 1 root root 974 Sep 22 1999 rusersd
-rwxr-xr-x 1 root root 941 Aug 16 1999 rwalld
-rwxr-xr-x 1 root root 882 Sep 9 1999 rwhod
-rwxr-xr-x 1 root root 1549 Sep 1 1999 sendmail
-rwxr-xr-x 1 root root 1451 Apr 15 1999 single
-rwxr-xr-x 1 root root 1177 Sep 25 1999 smb
-rwxr-xr-x 1 root root 851 Aug 31 1999 snmpd
-rwxr-xr-x 1 root root 2306 Sep 11 1999 squid
-rwxr-xr-x 1 root root 1027 Dec 21 14:04 syslog
-rwxr-xr-x 1 root root 1033 Jan 10 06:40 xfs
-rwxr-xr-x 1 root root 1212 Aug 16 1999 xntpd
-rwxr-xr-x 1 root root 1575 Sep 22 1999 ypbind
-rwxr-xr-x 1 root root 1084 Aug 17 1999 yppasswdd
-rwxr-xr-x 1 root root 1137 Aug 17 1999 ypserv
```

These services are can be functionally categorized as a system service or a network service. They are described in more detail in the section on Daemons and Services. For more information on how some of the script files for these services run, read the Linux startup Manual. Normally any of these services may be stopped, started, restarted, or status be checked by typing the name of one of these services (with the correct path) followed by the word stop, start, restart, or status respectively. For example, the line:

```
/etc/rc.d/init.d/nfs  restart
```

will restart network file sharing assuming it was running. To see the status type:

```
/etc/rc.d/init.d/nfs  status
```

The rc.local Script Program

The file "/etc/rc.d/rc3.d/S99.local" is a link file to the file "/etc/rc.d/rc.local". This file doesn't do much except for setting up the "/etc/issue" and "/etc/issue.net" files to reflect the system version when a user begins a terminal or telnet session. This is where most administrators will put any system customizations they want to make.

SUMMARY

1. Linuxconf Linuxconf is a extremely large project. It is tremendously powerful and has large number of features. Linuxconf is the BEST KNOWN system admin tool for Linux!! It is available in all distros like Redhat, Debian, Suse, and others.
2. The Linuxconf is the equivalent of SAM of HPUX and SMITTY of AIX. It runs on console window and as well as in GUI in X-Window. On Debian Linux to download and install just do 'apt-get linuxconf'.
3. The Webmin is the most powerful and very popular system admin tool for Linux. Webmin is so popular and it was awarded the best System Administration tool for Solaris, BSD, Linux, HPUX, AIX, SCO and others.
4. Webmin had been winning awards year after year for the past 5 years by many organisations, conferences and editors.
5. Webmin can also be used for admin of other OSes like AIX, Solaris, HPUX, IRIX, BSD, SCO Unix, OSF, Darwin, Apple Macintosh Mac OS X and Cygwin.
6. The best supported systems at the moment are Solaris, Linux (Redhat in particular) and FreeBSD.
7. Webmin is written in Perl and Perl is universal like "C" and unlike "C" it is scripting language. Perl is "cousin-brother" of "C" language.
8. But major drawback is that you need a X-Window system, Apache Webserver and Perl installed and running before using Webmin. And here is where Linuxconf fills the gap - you do not need X-Window and others to run Linuxconf.
9. Linuxconf can run even in dumb text-terminal console with ncurses.
10. Every file on your Linux system, including directories, is owned by a specific user and group. Therefore, file permissions are defined separately for users, groups, and others.

SHORT QUESTIONS WITH ANSWER

1. Create a new directory and create six pipes in that directory-

Ans.

```
[paul@rhel53 ~]$ mkdir pipes ; cd pipes
[paul@rhel53 pipes]$ mkfifo p1 p2 p3 p4 p5 p6
[paul@rhel53 pipes]$ ls -l
total 0
prw-rw-r-- 1  paul  paul  0  Apr   12 22:15  p1
prw-rw-r-- 1  paul  paul  0  Apr   12 22:15  p2
```

```
prw-rw-r-- 1 paul paul 0 Apr 12 22:15 p3
prw-rw-r-- 1 paul paul 0 Apr 12 22:15 p4
prw-rw-r-- 1 paul paul 0 Apr 12 22:15 p5
prw-rw-r-- 1 paul paul 0 Apr 12 22:15 p6
```

2. **Bounce a character between two pipes.**

Ans.
```
[paul@rhel53 pipes]$ echo -n x | cat - p1 > p2 &
[1] 4013
[paul@rhel53 pipes] $ cat <p2 =-pl &
[2] 4016
```

3. **Use top and ps to display information (pid, ppid, priority, nice value, ...) about these two cat processes.**

Ans.
```
top (probably the top two lines)
[paul@rhel53 pipes] $ ps -C cat
PID  TTY     TIME      CMD
4013 ptS/0   00:03:38  cat
4016 ptS/0   00:01:07  cat
[paul@rhel53 pipes) $ ps Cax | grep cat
4013 pts/0   R   4:00  |  \_ cat - pi
4016 pts/0   S   1:13  |  \_ cat
4044 pts/0   S+  0:00  |  \_ grep cat
```

4. **Bounce another character between two other pipes, but this time start the commands nice. Verify that all cat processes are battling for the cpu. (Feel free to fire up two more cats with the remaining; pipes).**

Ans.
```
echo -n y | nice cat - p3 > p4 &
nice cat <p4 >p3 &
```

5. **Use ps to verify that the two new cat processes have a nice value. Use the -o and -C options of ps for this.**

Ans.
```
[paul&rhel53 pipes) $ ps -C cat -o pid, ppid, pri., ni, comm
PID   PPID  PRI  NI  COMMAND
4013  3947  14   0   cat
4016  3947  21   0   cat
4025  3947  13   10  cat
4026  3947  13   10  cat
```

6. **Use renice te increase the nice value from 10 to 15. Notice the difference with the usual commands.**

Ans.
```
[paul@rhel53 pipes) $ renice +15 4025
4025: old priority 10, now priority 15
[paul@rhel53 pipes] $ renice +15 4026
4026; old priority 10, new priority 15
[paul@rhel53 pipes]$ ps -C cat -o pid, ppid, pri, ni, comm
PID   PPID  PRI  NI  COMMAND
4013  3947  14   0   cat
4016  3947  21   0   cat
4025  3947  9    15  cat
4026  3947  8    15  cat
```

1. Use the **jobs** command to verify whether you have any processes miming in background.
2. Use **vi** to create a little text file. Suspend **vi** in background.
3. Verify with **jobs** that **vi** is suspended in background.

4. Start **find / > allfiles.txt 2>/dev/null** in foreground. Suspend it in background before it finishes.
5. Start two long **sleep** processes in background.
6. Display all **jobs** in background.
7. Use the **kill** command to suspend the last **sleep** process.
8. Continue the **find** process in background (make sure it runs again).
9. Put one of the **sleep** commands back in foreground.
10. (if time permits, a general review question...) Explain in detail where the numbers come from in the next screenshot. When are the variables replaced by their value ? By which shell ?

```
[paul@RHEL4b ~]$ echo $$ $PPID
4224 4223
[paul@RHEL4b ~]$ bash -c "echo $$ $PPID"
4224 4223
[paul@RHEL4b ~]$ bash -c 'echo $$ $PPID'
5059 4224
[paul@RHEL4b ~]$ bash -c `echo $$ $PPID`
4223: 4224: command not found
```

EXERCISE

1. Which is the most important Part of UNIX Architecture?
2. Give different layers of UNIX Architecture. Explain the intended purposes of each.
3. What are the various services provided by an Operating System?
4. What is function of a Kernel?
5. Where does Kernel Resides in Memory?
6. How can the exceptions be resolved in UNIX?
7. What is the role of processor execution levels?
8. How are file organized in UNIX?
9. What is a Path, in context of Files & Directories?

❐❐❐

CHAPTER 3

MANAGING USER ACCOUNTS

3.0 INTRODUCTION

The user information is stored in the system /etc/passwd and /etc/shadow files, and that additionally, group membership information is stored in the /etc/group file. The update-passwd tool keeps the entries in these master files in sync on all Debian systems.While it is possible to edit these files by hand, it is not recommended.There exist several command line tools, which can be used to manage these files instead.

Now we are going to see the complete users and groups administration commands with examples and man pages.

Chapter Outline

3.1 PRINCIPLES

First we look at login logging ; how can we find out who is logging in to the system, when and from where. And who is not logging in, who fails at **su** or **ssh**.

Second we discuss how to configure the syslog daemon, and how to test it with **logger**.

3.2 MANAGING USER ACCOUNTS

When a new person joins an organisation he is usually given an account by the system administrator. This is the login account of the user. Now a days almost all Unix systems support an admin tool which seeks the following information from the system administrator to open a new account:

1. Username: This serves as the login name for the user.
2. Password: Usually a system administrator gives a simple password. The users are advised to later select a password which they feel comfortable

using. User's password appears in the shadow files in encrypted forms. Usually, the /etc/passwd file contains the information required by the login program to authenticate the login name and to initiate appropriate shell as shown in the description below:

```
bhatt:x:1007:1::/export/home/bhatt:/usr/local/bin/bash
damu:x:1001:10::/export/home/damu:/usr/local/bin/bash
```

Each line above contains information about one user. The first field is the name of the user; the next a dummy indicator of password, which is in another file, a shadow file. Password programs use a trap-door algorithm for encryption.

3. Home directory: Every new user has a home directory defined for him. This is the default login directory. Usually it is defined in the run command files.
4. Working set-up: The system administrators prepare .login and .profile files to help users to obtain an initial set-up for login. The administrator may prepare .cshrc, .xinitrc .mailrc .ircrc files. In Section 19.4 we shall later see how these files may be helpful in customizing a user's working environment. A natural point of curiosity would be: what happens when users log out? Unix systems receive signals when users log out. Recall, in Section 19.2 we mentioned that a user logs in under a login process initiated by *getty* process. Process *getty* identifies the terminal being used. So when a user logs out, the *getty* process which was running to communicate with that terminal is first killed. A new *getty* process is now launched to enable yet another prospective login from that terminal.

   The working set-up is completely determined by the startup files. These are basically .rc (run command) files. These files help to customize the user's working environment. For instance, a user's .cshrc file shall have a path variable which defines the access to various Unix built-in shell commands, utilities, libraries, etc.

   In fact, many other shell environmental variables like HOME, SHELL, MAIL, TZ (the time zone) are set up automatically. In addition, the .rc files define the access to network services or some need-based access to certain licensed software or databases as well. To that extent the *.rc* files help to customize the user's working environment. We shall discuss the role of run command files later in Section 19.4.
5. Group-id: The user login name is the user-id. Under Unix the access privileges are determined by the group a user belongs to. So a user is assigned a group-id. It is possible to obtain the id information by using an id command as shown below:

```
[bhatt@iiitbsun OS]$id
uid=1007(bhatt) gid=1(other)
[bhatt@iiitbsun OS]$
```

6. Disc quota: Usually a certain amount of disk space is allocated by default. In cases where the situation so warrants, a user may seek additional disk space. A user may interrogate the disk space available at any time by using the *df* command. Its usage is shown below:

```
df [options] [name] : to know the free disk space.
```

where name refers to a mounted file system, local or remote. We may specify directory if we need to know the information about that directory. The following options may help with additional information:

```
-l : for local file system
-t : reports total no. of allocated blocks and i-nodes on the device.
```

The Unix command *du* reports the number of disk blocks occupied by a file. Its usage is shown below:

du [options] [name]... where name is a directory or a file

Above name by default refers to the current directory. The following options may help with additional information:

-a : produce output line for each file

-s : report only the total usage for each name that is a directory, i.e. not individual files.

-r : produce messages for files that cannot be read or opened

7. Network services: Usually a user shall get a mail account. We will discuss the role of .mailrc file in this context in section 19.4. The user gets an access to Web services too.
8. Default terminal settings: Usually vt100 is the default terminal setting. One can attempt alternate terminal settings using *tset, stty, tput*, tabs with the control sequences defined in *terminfo termcap* with details recorded in /etc/ttytype or /etc/tty files and in shell variable TERM. Many of these details are discussed in Section 19.5.1 which specifically deals with terminal settings. The reader is encouraged to look up that section for details.

Once an account has been opened the user may do the following:

1. Change the password for access to one of his liking.
2. Customize many of the run command files to suit his needs.

Closing a user account

Here again the password file plays a role. Recall in section 19.1 we saw that /etc/password file has all the information about the users' home directory, password, shell, user and group-id, etc. When a user's account is to be deleted, all of this information needs to be erased. System administrators login as root and delete the user entry from the password file to delete the account.

3.3 PASSWORD FILE

The login program will prompt for the user name if no argument is given on the command line.

If the file "/etc/nologin" exists and the user is not root, the contents of the "/etc/nologin" file are printed to the screen and the login is terminated. If special access restrictions are specified for the user logging in in the file "etc/usertty", the restrictions must be met or the log in will be denied and the program syslog will log the attempt. If the user is root the login must be on a terminal listed in the file "etc/securetty".

If the above conditions are met, the user password will be requested and then it will be checked (If a password is required for this username). After three unsuccessful attempts to login the response gets very slow, and after 10 attempts, login dies. As usual all login failures will be reported by the syslog facility. If the file ".hushlogin" exists in the user's home directory then a "quiet" login is performed which disables checking of mail and the printing of the last login time and the message of the day. Otherwise if the file "var/log/lastlog" exists the last login time is printed and then the current login is recorded in this file. Is the current login recorded in this file if it does not already exist or if the file ".hushlogin" exists? I think it does but have found no documentation that says.

At this point the login program will perform standard administrative tasks. These include:

1. Setting the UID and GID of the tty
2. Preserving the TERM environment variable if it exists.

3. Preserving other environment variables if the –p option is used
4. The HOME, PATH, SHELL, TERM, MAIL, and LOGNAME environment variables are set.
5. The default path is set to "/usr/local/bin:/bin:/usr/bin:." for normal users and "/sbin:/bin:/usr/sbin" for root.
6. If this is not a "quiet" login, the message of the day is printed and the file with the user's name in "/usr/spool/mail" will be checked and a message will be printed if it has non-zero length.
7. The users shell is started. The shell is specified in the file "/etc/passwd". If it is not specified, login will use "/bin/sh" as a default shell. This shell will be run with the user's privileges rather than root privileges as login was run.
8. If there is no directory specified for the user in "/etc/passwd", login will use "/" by default for the user's home directory.

Another function that login will perform is to update the user accounting login files which are "/var/run/utmp" and "var/log/wtmp" which hold information about the amount of time users have been on the system along with when they logged on and off. Also the init program and getty may write to these files.

How login uses the /etc/passwd file:

Once the user has successfully logged in, the login program will invoke the user's shell. The login program will look in the /etc/passwd file to determine which shell program to run. The /etc/passwd file contains entries containing the complete path of the shell. A sample /etc/passwd file is listed below:

```
root:x:0:0:root:/root:/bin/bash
bin:x:1:1:bin:/bin:
daemon:x:2:2:daemon:/sbin:
adm:x:3:4:adm:/var/adm:
lp:x:4:7:lp:/var/spool/lpd:
sync:x:5:0:sync:/sbin:/bin/sync
shutdown:x:6:0:shutdown:/sbin:/sbin/shutdown
halt:x:7:0:halt:/sbin:/sbin/halt
mail:x:8:12:mail:/var/spool/mail:
news:x:9:13:news:/var/spool/news:
uucp:x:10:14:uucp:/var/spool/uucp:
operator:x:11:0:operator:/root:
games:x:12:100:games:/usr/games:
gopher:x:13:30:gopher:/usr/lib/gopher-data:
ftp:x:14:50:FTP User:/home/ftp:
nobody:x:99:99:Nobody:/:
xfs:x:100:101:X Font Server:/etc/X11/fs:/bin/false
gdm:x:42:42::/home/gdm:/bin/bash
postgres:x:40:233:PostgreSQL Server:/var/lib/pgsql:/bin/bash
squid:x:23:23::/var/spool/squid:/dev/null
mark:x:500:500::/home/mark:/bin/bash
george:x:501:501::/home/george:/bin/bash
```

the syntax is:

account:password:UID,GID,GECOS:directory:shell

where the fields are defined as:

- account - The user's name.
- password - The users encrypted passwrod or a place holding character if the system is using shadow passwords and storing the password in the /etc/shadow file which is readable only by root.
- UID - The users numerical identification.
- GID - The number of the primary group for the user.
- GECOS - Usually has the full user name. This field is only for information purposes and is optional. This information is sometimes called the user's finger information.
- directory - The full path of the user's home directory.
- shell - The full path and filename of the user's shell. If no value is here /bin/sh is assumed. This value can be changed with the chsh command.

The login program will use the account field to find the username and therefore get the UID of the user. Login will also use the password (or the /etc/shadow file) to be sure the entered password is a match. Login will look up the user's home directory and use that to set the $HOME environment variable. Login will use the shell field to determine what shell program (such as bash, sh, tsh, etc) to run for that user. then login will pass program control to the shell program. There is an important difference in the control passed at this point, however! The shell program will run with the user's privileges and not with root privileges. The programs to this point (init, getty, login) have all run with root privileges.

Files used by the login program:

- /etc/nologin - This file is used to prevent users who are not root from logging into the system.
- /etc/usertty - This file is used to impose special access restrictions on users.
- /etc/securetty - Controls the terminals that the root user can login on.
- .hushlogin - When this file exists in the user's home directory, it will prevent check for mail, printing of the last login time, and the message of the day when the user logs in.
- /var/log/lastlog - Contains information about the last time a login was done on the system.
- /etc/passwd - Contains information about the user including the ID, name, home directory, and the path to the preferred shell program. If not using shadow passwords, this file may also contain user passwords.

To keep track of who is logging into the system, Linux can maintain the **/var/log/wtmp**, **/var/log/btmp**, **/var/run/utmp** and **/var/log/lastlog** files.

/var/run/utmp (who)

Use the **who** command to see the /var/run/utmp file. This command is showing you all the **currently** logged in users. Notice that the utmp file is in /var/run and not in /var/log .

```
[root@rhel4 ~]# who
paul pts/1 Feb 14 18:21 (192.168.1.45)
sandra pts/2 Feb 14 18:11 (192.168.1.42)
```

```
inge pts/3 Feb 14 12:01 (192.168.1.33)
els pts/4 Feb 14 14:33 (192.168.1.19)
```

/var/log/wtmp (last)

The /var/log/wtmp file is updated by the **login program**. Use **last** to see the /var/run/wtmp file.

```
[root@rhel4a ~]# last | head
paul pts/1 192.168.1.45 Wed Feb 14 18:39 still logged in
reboot system boot 2.6.9-42.0.8.ELs Wed Feb 14 18:21 (01:15)
nicolas pts/5 pc-dss.telematic Wed Feb 14 12:32 - 13:06 (00:33)
stefaan pts/3 pc-sde.telematic Wed Feb 14 12:28 - 12:40 (00:12)
nicolas pts/3 pc-nae.telematic Wed Feb 14 11:36 - 12:21 (00:45)
nicolas pts/3 pc-nae.telematic Wed Feb 14 11:34 - 11:36 (00:01)
dirk pts/5 pc-dss.telematic Wed Feb 14 10:03 - 12:31 (02:28)
nicolas pts/3 pc-nae.telematic Wed Feb 14 09:45 - 11:34 (01:48)
dimitri pts/5 rhel4 Wed Feb 14 07:57 - 08:38 (00:40)
stefaan pts/4 pc-sde.telematic Wed Feb 14 07:16 - down (05:50)
[root@rhel4a ~]#
```

The last command can also be used to get a list of last reboots.

```
[paul@rekkie ~]$ last reboot
reboot system boot 2.6.16-rekkie Mon Jul 30 05:13 (370+08:42)
wtmp begins Tue May 30 23:11:45 2011
[paul@rekkie ~]$
```

/var/log/lastlog (lastlog)

Use **lastlog** to see the /var/log/lastlog file.

```
[root@rhel4a ~]# lastlog | tail
tim pts/5 10.170.1.122 Tue Feb 13 09:36:54 +0100 2012
rm pts/6 rhel4 Tue Feb 13 10:06:56 +0100 2012
henk **Never logged in**
stefaan pts/3 pc-sde.telematic Wed Feb 14 12:28:38 +0100 2012
dirk pts/5 pc-dss.telematic Wed Feb 14 10:03:11 +0100 2012
arsene **Never logged in**
nicolas pts/5 pc-dss.telematic Wed Feb 14 12:32:18 +0100 2012
dimitri pts/5 rhel4 Wed Feb 14 07:57:19 +0100 2012
bashuserrm pts/7 rhel4 Tue Feb 13 10:35:40 +0100 2012
kornuserrm pts/5 rhel4 Tue Feb 13 10:06:17 +0100 2012
[root@rhel4a ~]#
```

/var/log/btmp (lastb)

There is also the **lastb** command to display the **/var/log/btmp** file. This file is updated by the login program when entering the wrong password, so it contains failed login attempts. Many computers will not have this file, resulting in no logging of failed login attempts.

```
[root@RHEL4b ~]# lastb
lastb: /var/log/btmp: No such file or directory
Perhaps this file was removed by the operator to prevent logging
lastb\info.
```

```
[root@RHEL4b ~]#
```

The reason given for this is that users sometimes type their password by mistake instead of their login, so this world readable file poses a security risk. You can enable bad login logging by simply creating the file. Doing a chmod o-r /var/log/btmp improves security.

```
[root@RHEL4b ~]# touch /var/log/btmp
[root@RHEL4b ~]# ll /var/log/btmp
-rw-r-r- 1 root root 0 Jul 30 06:12 /var/log/btmp
[root@RHEL4b ~]# chmod o-r /var/log/btmp
[root@RHEL4b ~]# lastb
btmp begins Mon Jul 30 06:12:19 2012
[root@RHEL4b ~]#
```

Failed logins via ssh, rlogin or su are not registered in /var/log/btmp. Failed logins via tty are.

```
[root@RHEL4b ~]# lastb
HalvarFl tty3 Mon Jul 30 07:10 - 07:10 (00:00)
Maria tty1 Mon Jul 30 07:09 - 07:09 (00:00)
Roberto tty1 Mon Jul 30 07:09 - 07:09 (00:00)
btmp begins Mon Jul 30 07:09:32 2012
[root@RHEL4b ~]#
```

3.4 PASSWORD SECURITY

su and ssh logins

Depending on the distribution, you may also have the **/var/log/secure** file being filled with messages from the auth and/or authpriv syslog facilities. This log will include su and/or ssh failed login attempts. Some distributions put this in **/var/log/auth.log**, verify the syslog configuration.

```
[root@RHEL4b ~]# cat /var/log/secure
Jul 30 07:09:03 sshd[4387]: Accepted publickey for paul from ::ffff:19\
2.168.1.52 port 33188 ssh2
Jul 30 05:09:03 sshd[4388]: Accepted publickey for paul from ::ffff:19\
2.168.1.52 port 33188 ssh2
Jul 30 07:22:27 sshd[4655]: Failed password for Hermione from ::ffff:1\
92.168.1.52 port 38752 ssh2
Jul 30 05:22:27 sshd[4656]: Failed password for Hermione from ::ffff:1\
92.168.1.52 port 38752 ssh2
Jul 30 07:22:30 sshd[4655]: Failed password for Hermione from ::ffff:1\
92.168.1.52 port 38752 ssh2
Jul 30 05:22:30 sshd[4656]: Failed password for Hermione from ::ffff:1\
92.168.1.52 port 38752 ssh2
Jul 30 07:22:33 sshd[4655]: Failed password for Hermione from ::ffff:1\
92.168.1.52 port 38752 ssh2
Jul 30 05:22:33 sshd[4656]: Failed password for Hermione from ::ffff:1\
92.168.1.52 port 38752 ssh2
Jul 30 08:27:33 sshd[5018]: Invalid user roberto from ::ffff:192.168.1\
.52
```

```
Jul 30 06:27:33 sshd[5019]: input_userauth_request: invalid user rober\to
Jul 30 06:27:33 sshd[5019]: Failed none for invalid user roberto from\
::ffff:192.168.1.52 port 41064 ssh2
Jul 30 06:27:33 sshd[5019]: Failed publickey for invalid user roberto\
from ::ffff:192.168.1.52 port 41064 ssh2
Jul 30 08:27:36 sshd[5018]: Failed password for invalid user roberto f\
rom ::ffff:192.168.1.52 port 41064 ssh2
Jul 30 06:27:36 sshd[5019]: Failed password for invalid user roberto f\
rom ::ffff:192.168.1.52 port 41064 ssh2
[root@RHEL4b ~]#
```

You can enable this yourself, with a custom log file by adding the following line tot syslog.conf.

```
auth.*,authpriv.*  /var/log/customsec.log
```

3.5 USERS AND GROUPS

There are several characteristics of passwords and how they are or should be stored on your system. They should be:

1. In a file that is readable only by root.
2. In a one way hash format.

The actual password is not stored on your system, but its one way hash value is.

3.5.1 One way hash

A one way hash is a function. The password is given to the function, and the function generates an output. This function is special since it has the following characteristics: 1. It is easy to generate a output from it that will produce the same output for the same password. 2.It is very difficult to generate the original input from the produced output. An example would be like a sine to inverse sine function, square to square root, etc. There are one way hash protocols that are used for this purpose. One popular protocol used on Linux is called MD5. Therefore the way "crack" password cracking programs work is to get a copy of your system password file and try to guess at the password, running it through the one way hash function until it gets a match. This is why short passwords are not good. There is another solution to help this situation called salting, but I'm not sure it is implemented yet on Linux. I think it may be, but is only a two character string, and its value may be stored in /etc/password, which is world readable so it may not be effective in increasing protection against any of your users. Basically, when a short password is received it is "salted" with extra characters at random. The extra salt characters must be stored somewhere on the system in a secure location and used to generate the hashed password value for comparison to the stored value anytime the user logs in. The "longer" salted password and the associated output will make it more difficult for crack programs to guess the original password.

3.5.2 Protecting Your Hashed Password Value

In the past, the user's password was stored in a file called "/etc/passwd". This file needs to be readable by all users, however, since programs such as the shell that run at the user's privilege level need to access information in it such as the user's name. A normal listing of the /etc/passwd file shows the user access settings:

```
-rw-r-r- 1 root root 775 May 5 12:43 /etc/passwd
```

This can cause a problem compromising system security. This is why shadow passwords was implemented. See the section on the login process for a description of the /etc/passwd file and how it is used.

3.6 SHADOW FILE

Traditional Unix systems keep user account information, including one-way encrypted passwords, in a text file called "/etc/passwd". As this file is used by many tools (such as "ls") to display file ownerships, etc. by matching user id #'s with the user's names, the file needs to be world-readable. Consequentally, this can be somewhat of a security risk.

Another method of storing account information, one that I always use, is with the shadow password format. As with the traditional method, this method stores account information in the /etc/passwd file in a compatible format. However, the password is stored as a single "x" character (i.e., not actually stored in this file). A second file, called ``/etc/shadow", contains encrypted password as well as other information such as account or password expiration values, etc. The /etc/shadow file is readable only by the root account and is therefore less of a security risk.

While some other Linux distributions forces you to install the Shadow Password Suite in order to use the shadow format, Red Hat makes it simple. To switch between the two formats, type (as root):

```
 /usr/sbin/pwconv To convert to the shadow format
/usr/sbin/pwunconv To convert back to the traditional format
```

With shadow passwords, the "/etc/passwd" file contains account information, and looks like this:

```
smithj:x:561:561:Joe Smith:/home/smithj:/bin/bash
```

Each field in a passwd entry is separated with ":" colon characters, and are as follows:

- Username, up to 8 characters. Case-sensitive, usually all lowercase
- An "x" in the password field. Passwords are stored in the ``/etc/shadow" file.
- Numeric user id. This is assigned by the "adduser" script. Unix uses this field, plus the following group field, to identify which files belong to the user.
- Numeric group id. Red Hat uses group id's in a fairly unique manner for enhanced file security. Usually the group id will match the user id.
- Full name of user. I'm not sure what the maximum length for this field is, but try to keep it reasonable (under 30 characters).
- User's home directory. Usually /home/username (e.g. /home/smithj). All user's personal files, web pages, mail forwarding, etc. will be stored here.
- User's "shell account". Often set to ``/bin/bash" to provide access to the bash shell (my personal favorite shell).

Perhaps you do not wish to provide shell accounts for your users. You could create a script file called "/bin/sorrysh", for example, that would display some kind of error message and log the user off, and then set this script as their default shell.

NOTE: If the account needs to provide "FTP" transfers to update web pages, etc. then the shell account will need to be set to ``/bin/bash" — and then special permissions will need to be set up in the user's home directory to prevent shell logins.

The "/etc/shadow" file contains password and account expiration information for users, and looks like this:

```
smithj:Ep6mckrOLChF.:10063:0:99999:7:::
```

3.7 GROUPS AND THE GROUP FILE

The file's permissions, divided into three groups, each consisting of three characters. The first group of three characters shows the read, write, and execute permissions for *user*, the owner of the file. The next group shows the read, write, and execute permissions for the *group* of the file. Similarly, the last group of three characters shows the permissions for *other*, everyone else. In each group, the first character means the read permission, the second one write permission, and the third one execute permission.

The characters are pretty easy to remember.

r = read permission
w = write permission
x = execute permission
- = no permission

What does this mean in practice? Let's have an example. Remember the imaginary directory listing we did at the beginning? The output looked like this:

```
drwxr-xr-x  3  nana  writers  80  2012-09-20  21:37  dir
-rw-r--  1  nana  writers  8187  2012-09-19  13:35  file
-rwxr-xr-x  1  nana  writers  10348  2012-07-17  20:31  otherfile
```

As we already noticed, dir is a directory, because the first column begins with a d. The owner of this directory is user nana and the group owner is writers. The first three characters, rwx, indicate the directory's owner, nana in this case, has full access to the directory. The user nana is able to access, view, and modify the files in that directory. The next three characters, r-x, indicate that all users belonging to group writers have read and execute permissions to the directory. They can change into the directory, execute files, and view its contents. However, because they don't have write permissions, they can't make any changes to the directory content. Finally, the last three characters, r-x, indicate that all the users who are not nana or don't belong into group writers, have read and execute permissions in the directory.

How about file? Because the first column begins with a -, the file is a regular file, owned by user nana and group writers, just like the directory in our example. The first three characters, rw-, indicate the owner has read and write access to the file. According to the next three characters, r—, the users belonging to group writers can view the file but not modify or execute it. The final three characters, —-, indicate no one else has any access to the file.

Similarly, you can see otherfile is a regular file and its owner has full access to it, while everyone else can read and execute the file but not modify it.

< How to set file permissions - symbolic mode >

You can set file permissions with the chmod command. Both the root user and the file's owner can set file permissions. chmod has two modes, symbolic and numeric.

The symbolic mode is pretty easy to remember. First, you decide if you set permissions for the user (u), the group (g), others (o), or all of the three (a). Then, you either add a permission (+), remove it (-), or wipe out the previous permissions and add a new one (=). Next, you decide if you set the read permission (r), write permission (w), or execute permission (x). Last, you'll tell chmod which file's permissions you want to change.

Let's have a couple of examples. Suppose we have a regular file called testfile, and the file has full access permissions for all the groups (long directory listing would show -rwxrwxrwx as the file's permissions).

Wipe out all the permissions but add read permission for everybody:

$ chmod a=r testfile

After the command, the file's permissions would be -r—r—r—

Add execute permissions for group:

$ chmod g+x testfile

Now, the file's permissions would be -r—r-xr—

Add both write and execute permissions for the file's owner. Note how you can set more than one permission at the same time:

$ chmod u+wx testfile

After this, the file permissions will be -rwxr-xr—

Remove the execute permission from both the file's owner and group. Note, again, how you can set them both at once:

$ chmod ug-x testfile

Now, the permissions are -rw-r—r—

As a summary, have a look at this quick reference for setting file permissions in symbolic mode:

Which user?	
u	user/owner
g	group
o	other
a	all

What to do?	
+	add this permission
-	remove this permission
=	set exactly this permission

Which permissions?	
r	read
w	write
x	execute

< How to set file permissions - numeric mode >

The other mode in which chmod can be used is the numeric mode. In the numeric mode, the file permissions aren't represented by characters. Instead, they are represented by a three-digit octal number.

4 = read (r)
2 = write (w)
1 = execute (x)
0 = no permission (-)

To get the permission bits you want, you add up the numbers accordingly. For example, the rwx permissions would be 4+2+1=7, rx would be 4+1=5, and rw would be 4+2=6. Because you set separate permissions for the owner, group, and others, you'll need a three-digit number representing the permissions of all these groups.

Let's have an example.

$ chmod 755 testfile

This would change the testfile's permissions to -rwxr-xr-x. The owner would have full read, write, and execute permissions (7=4+2+1), the group would have read and execute permissions (5=4+1), and the world would have the read and execute permissions as well.

Let's have another example:

$ chmod 640 testfile

In this case, testfile's permissions would be -rw-r——-. The owner would have read and write permissions (6=4+2), the group would have read permissions only (4), and the others wouldn't have any access permissions (0).

The numeric mode may not be as straightforward as the symbolic mode, but with the numeric mode, you can more quickly and efficiently set the file permissions. This quick reference for setting file permissions in numeric mode might help:

Which number?	
0	—
1	—x
2	-w-
3	-wx
4	r—
5	r-x
6	rw-
7	rwx

3.8 SHELLS AND RESTRICTED SHELLS

In FreeBSD, a lot of everyday work is done in a command line interface called a shell. A shell's main job is to take commands from the input channel and execute them. A lot of shells also have built in functions to help with everyday tasks such as file management, file globbing, command line editing, command macros, and environment variables. FreeBSD comes with a set of shells, such as sh, the Bourne Shell, andtcsh, the improved C-shell. Many other shells are available from the FreeBSD Ports Collection, such as zsh and bash.

Which shell do you use? It is really a matter of taste. If you are a C programmer you might feel more comfortable with a C-like shell such as tcsh. If you have come from Linux or are new to a UNIX® command line interface you might try bash. The point is that each shell has unique properties that may or may not work with your preferred working environment, and that you have a choice of what shell to use.

One common feature in a shell is filename completion. Given the typing of the first few letters of a command or filename, you can usually have the shell automatically complete the rest of the command or filename by hitting the **Tab** key on the keyboard. Here is an example. Suppose you have two files called foobar and foo.bar. You want to deletefoo.bar. So, what you would type on the keyboard is: rm fo[**Tab**].[**Tab**].

The shell would print out rm foo[BEEP].bar.

The [BEEP] is the console bell, which is the shell telling me it was unable to totally complete the filename because there is more than one match. Both foobar and foo.bar start with fo, but it was able to complete to foo. If you type in ., then hit **Tab** again, the shell would be able to fill in the rest of the filename for you.

Another feature of the shell is the use of environment variables. Environment variables are a variable/key pair stored in the shell's environment space. This space can be read by any program invoked by the shell, and thus contains a lot of program configuration. Here is a list of common environment variables and what they mean:

Variable	Description
USER	Current logged in user's name.
PATH	Colon-separated list of directories to search for binaries.
DISPLAY	Network name of the X11 display to connect to, if available.
SHELL	The current shell.
TERM	The name of the user's type of terminal. Used to determine the capabilities of the terminal.
TERMCAP	Database entry of the terminal escape codes to perform various terminal functions.
OSTYPE	Type of operating system. e.g., FreeBSD.
MACHTYPE	The CPU architecture that the system is running on.
EDITOR	The user's preferred text editor.
PAGER	The user's preferred text pager.
MANPATH	Colon-separated list of directories to search for manual pages.

Setting an environment variable differs somewhat from shell to shell. For example, in the C-Style shells such as tcsh and csh, you would use setenv to set environment variables. Under Bourne shells such as sh and bash, you would use export to set your current environment variables. For example, to set or modify the EDITOR environment variable, under csh or tcsh a command like this would set EDITOR to/usr/local/bin/emacs:

```
% setenv EDITOR /usr/local/bin/emacs
```

Under Bourne shells:

```
% export EDITOR="/usr/local/bin/emacs"
```

You can also make most shells expand the environment variable by placing a $ character in front of it on the command line. For example, echo $TERM would print out whatever $TERM is set to, because the shell expands $TERM and passes it on to echo.

Shells treat a lot of special characters, called meta-characters as special representations of data. The most common one is the * character, which represents any number of characters in a filename. These special meta-characters can be used to do filename globbing. For example, typing in echo * is almost the same as typing in ls because the shell takes all the files that match * and puts them on the command line for echo to see.

To prevent the shell from interpreting these special characters, they can be escaped from the shell by putting a backslash (\) character in front of them. echo $TERM prints whatever your terminal is set to. echo \$TERM prints $TERM as is.

3.8.1 Changing Your Shell

The easiest way to change your shell is to use the chshcommand. Running chsh will place you into the editor that is in your EDITOR environment variable; if it is not set, you will be placed in vi. Change the "Shell:" line accordingly.

You can also give chsh the -s option; this will set your shell for you, without requiring you to enter an editor. For example, if you wanted to change your shell to bash, the following should do the trick:

```
% chsh -s /usr/local/bin/bash
```

Note: The shell that you wish to use *must* be present in the /etc/shells file. If you have installed a shell from the ports collection, then this should have been done for you already. If you installed the shell by hand, you must do this.

For example, if you installed bash by hand and placed it into /usr/local/bin, you would want to:

```
# echo "/usr/local/bin/bash" >> /etc/shells
```

Then rerun chsh

You can use the passwd command to change user or group accounts password. A normal user may only change the password for his/her own account, the super user (root) may change the password for any account. You can use the passwd command for locking or unlocking an account.

3.9 RESTRICTED SHELLS

The Restricted Shell

If Bash is started with the name rbash, or the —restricted or -r option is supplied at invocation, the shell becomes restricted. A restricted shell is used to set up an environment more controlled, than the standard shell. A restricted shell behaves identically to bash with the exception that the following are disallowed or not performed:

- Changing directories with the cd builtin.
- Setting or unsetting the values of the SHELL, PATH, ENV, or BASH_ENV variables.
- Specifying command names containing slashes.
- Specifying a filename containing a slash as an argument to the . builtin command.
- Specifying a filename containing a slash as an argument to the -p option to the hash builtin command.
- Importing function definitions from the shell environment at startup.
- Parsing the value of SHELLOPTS from the shell environment at startup.
- Redirecting output using the '>', '>|', '<>', '>&', '&>', and '>>' redirection operators.
- Using the exec builtin to replace the shell with another command.
- Adding or deleting builtin commands with the -f and -d options to the enable builtin.

- Using the enable builtin command to enable disabled shell builtins.
- Specifying the -p option to the command builtin.
- Turning off restricted mode with 'set +r' or 'set +o restricted'.

These restrictions are enforced after any startup files are read.

Disabled commands in restricted shells

- Running a script or portion of a script in *restricted mode* disables certain commands that would otherwise be available. This is a security measure intended to limit the privileges of the script user and to minimize possible damage from running the script. The following commands and actions are disabled:
- Using *cd* to change the working directory.
- Changing the values of the *$PATH*, *$SHELL*, *$BASH_ENV*, or *$ENV* environmental variables.
- Reading or changing the *$SHELLOPTS*, shell environmental options.
- Output redirection.
- Invoking commands containing one or more /'s.
- Invoking exec to substitute a different process for the shell.
- Various other commands that would enable monkeying with or attempting to subvert the script for an unintended purpose.
- Getting out of restricted mode within the script.

EXAMPLE 3.1. *Running a script in restricted mode*

```
#!/bin/bash

# Starting the script with "#!/bin/bash -r"
#+ runs entire script in restricted mode.

echo

echo "Changing directory."
cd /usr/local
echo "Now in `pwd`"
echo "Coming back home."
cd
echo "Now in `pwd`"
echo

# Everything up to here in normal, unrestricted mode.

set -r
# set -restricted has same effect.
echo "==> Now in restricted mode. <=="

echo
echo

echo "Attempting directory change in restricted mode."
cd ..
echo "Still in `pwd`"
```

```
echo
echo

echo "\$SHELL = $SHELL"
echo "Attempting to change shell in restricted mode."
SHELL="/bin/ash"
echo
echo "\$SHELL= $SHELL"
echo
echo

echo "Attempting to redirect output in restricted mode."
ls -l /usr/bin > bin.files
ls -l bin.files # Try to list attempted file creation effort.

echo

exit 0
```

3.10 USER MANAGEMENT COMMANDS

Add New User in Linux

useradd - Create a new user or update default new user information

Syntax

```
useradd [-c comment] [-d home_dir] [-e expire_date] [-f inactive_days] [-
g initial_group] [-G group[,...]] [-m [-k skeleton_dir]] [-o] [-p passwd]
[-s shell] [-u uid] login

useradd -D [-g default_group] [-b default_home] [-e default_expire_date]
[-f default_inactive] [-s default_shell]
```

If you want to know more available options you need to check the useradd man page

Examples

Adding New User

First you need to create three (test1,test2,admin1) groups for our examples using groupadd

Options

-d home directory
-s starting program (shell)
-p password
-g (primary group assigned to the users)
-G (Other groups the user belongs to)
-m (Create the user's home directory)
To add a new user with
a primary group of test1
a second group test2
starting shell /bin/bash

password of xxxx
home directory of admin
create home directory
a login name of admin
#useradd -g test1 -G test2 -s /bin/bash -p xxxx -d/home/admin -m admin
This will create a new user admin.

One additional switch worth mentioning is "-D", which controls the defaults for useradd.

Specifying the "-D" switch on its own will simply display the default settings, while specifying -D in conjunction with other switches will change the defaults to those values.

```
# useradd -D
GROUP=100
INACTIVE=-1HOME=/home
INACTIVE=-1
EXPIRE=
SHELL=/bin/bash
SKEL=/etc/skel
# useradd -D -s /bin/sh
# useradd -D
GROUP=100
HOME=/home
INACTIVE=-1
EXPIRE=
SHELL=/bin/sh
SKEL=/etc/skel
```

As you can see, this changes the default shell for created users from "bash" to "sh".

adduser - User Friendly Frontend for useradd command

Syntax

adduser [options] user group

If you want to know available option refer add user man page

Example

```
#adduser admin
Adding user 'admin' ...
Adding new group 'admin' (1001) ...
Adding new user 'admin' (1001) with group 'admin' ...
Creating home directory '/home/admin' ...
Copying files from '/etc/skel' ...
Enter new UNIX password:
Retype new UNIX password:
passwd: password updated successfully
Changing the user information for admin
Enter the new value, or press ENTER for the default
Full Name []: Admin
Room Number []:
```

Work Phone []:
Home Phone []:
Other []:
Is the information correct? [y/N] y

You'll notice that, by default, the adduser command creates a group with the same name as the username, and makes this group the primary group for that user. This is called a user private group (UPG)

Modify User in Linux

usermod - Modify a user account

Syntax

usermod [-c comment] [-d home_dir [-m]] [-e expire_date] [-f inactive_days] [-g initial_group] [-G group [,...]] [-l login_name] [-p passwd] [-s shell] [-u uid [-o]] [-L|-U] login

If you want to know available option check usermod man page

Example

Options:
-d home directory
-s starting program (shell)
-p password
-g (primary group assigned to the users)
-G (Other groups the user belongs to)
To add the group 'others' to the user admin
#usermod -G others admin

Delete User in Linux

userdel - Delete a user account and related files

Syntax

userdel [-r] login
If you want to know available options check userdel man page

Example

Options
-r (remove home directory)
To remove the user 'admin' and his home directory
#userdel -r admin

deluser - remove a user from the system

Syntax

deluser [options] user group
If you want more options check deluser man page

Example

By default, deluser will remove the user without removing the home directory, the mail spool or any other files on the system owned by the user. Removing the home directory and mail spool can be achieved using the —remove-home option. If the —home option is given, deluser will only remove the user if the directory given to the —home option matches the user's real home directory.

#deluser —remove-home admin

Groups Administration in Linux

Add New Group in Linux

groupadd - Create a new group

Syntax

groupadd [-g gid [-o]] group

For more options check groupadd man page

Example

#groupadd test1

This will create a test1 group

addgroup - add a group to the system

Syntax

addgroup [options] [—gid ID] group

If you want to know available options check addgroup man page

#addgroup

Enter a groupname to add: admin1

Adding group 'admin12†(1001)...Done.

Modify Group in Linux

groupmod - Modify a group

Syntax

groupmod [-g gid [-o]] [-n group_name] group

For more options check groupmod man page

Example

#groupmod test1 test2

This will modify group name test1 to test2

Delete group in Linux

groupdel - Delete a group

Syntax

groupdel groupname

For more options check groupdel man page

Example

#groupdel test2

this will delete the test2 group

delgroup - remove a group from the system

Syntax

delgroup [options] [—only-if-empty] group

For more details about options check delgroup man page

Example

#delgroup —only-if-empty test2

Removing group 'test2 ...done.

groups Command

print the groups a user is in

Syntax

groups [username]

This simple command displays what groups a user is a member of. It takes the username of user as a parameter. If no username is given, it defaults to the current user.

groups
root
groups admin
test1 : test2

3.11 HOMES, PERMISSIONS AND DEFAULT FILES

Ownership

Every file in UNIX has an **owner user** and an **owner group**. So, for any file in the system, user 'nick' may have one of the following ownership relations:

- nick **owns the file**, i.e. the file's owner is 'nick'.
- nick is **a member of the group that owns the file**, i.e. the file's owner group is 'perlfect'.
- nick is neither the owner, nor belonging to the group that owns the file.

Permissions

Every file on the system has associated with it a set of permissions. Permissions tell UNIX what can be done with that file and by whom. There are three things you can (or can't) do with a given file:

- **read** it,
- **write** (modify) it and
- **execute** it.

Unix permissions specify which of the above operations can be performed for any ownership relation with respect to the file. In simpler terms, what can the owner do, what can the owner group do, and what can everybody else do with the file. For any given ownership relation, we need three bits to specify access permissions: the first to denote read (r) access, the second to denote (w) access and the third to denote execute (x) access. We have three ownership relations: 'owner', 'group' and 'all' so we need a triplet for each, resulting in nine bits. Each bit can be *set* or *clear*. (not set) We mark a set bit by it's corresponding operation letter (r, w or x) and a clear bit by a dash (-) and put them all on a row. An example might be rwxr-xr-x.What this means is that the owner can do anything with the file, but group owners and the rest of the world can only read or execute it. Usually in UNIX there is also another bit that precedes this 9-bit pattern. You do not need to know about it, at least for the time being.

So, if you try ls -l on the command prompt you will get something like the following:

```
[nick@thekla src]$ ls -l
-rwxr-xr-x    1 nick   users 382    Jan    19 11:49    bscoped.pl
drwxr-xr-x    3 nick   users 1024   Jan    19 11:19    lib/
-rwxr-xr-x    1 nick   users 1874   Jan    19 10:23    socktest.pl
```

The first column here shows the permission bit pattern for each file. The third column shows the owner, and the fourth column shows the owner group. By the time, the information provided by ls -l should be enough for you to figure out what each user of the system can do with any of the files in the directory.

Directories

Another interesting thing to note is that lib/ which is a directory has permissions, too. Permissions take a different meaning for directories. Here's what they mean:

- read determines if a user can view the directory's contents, i.e. do ls in it.
- write determines if a user can create new files or delete file in the directory. (Note here that this essentially means that a user with write access toa directory can delete files in the directory *even* if he/she doesn't have write permissions for the file! So be careful with this.)
- execute determines if the user can cd into the directory.

chmod

To set/modify a file's permissions you need to use the chmod program. Of course, only the owner of a file may use chmod to alter a file's permissions. chmod has the following syntax: **chmod [options] mode file(s)**

The 'mode' part specifies the new permissions for the file(s) that follow as arguments. A mode specifies which user's permissions should be changed, and afterwards which access types should be changed. Let's say for example: chmod a-x socktest.pl This means that the e**x**ecute bit should be cleared (-) for **a**ll users. (owner, group and the rest of the world) The permissions start with a letter specifying what users should be affected by the change, this might be any of the following:

- **u** the owner user
- **g** the owner group
- **o** others (neither u, nor g)
- **a** all users

This is followed by a change instruction which consists of a +(set bit) or -(clear bit) and the letter corresponding to the bit that should be changed.

Let's see some examples:

```
$ ls -l socktest.pl
-rwxr-xr-x 1 nick users 1874 Jan 19 10:23 socktest.pl*

$ chmod a-x socktest.pl
$ ls -l socktest.pl
-rw-r-r- 1 nick users 1874 Jan 19 10:23 socktest.pl

$ chmod g+w socktest.pl
$ ls -l socktest.pl
-rw-rw-r- 1 nick users 1874 Jan 19 10:23 socktest.pl
```

```
$ chmod ug+x socktest.pl
$ ls -l socktest.pl
-rwxrwxr- 1 nick users 1874 Jan 19 10:23 socktest.pl*

$ chmod ug-wx socktest.pl
$ ls -l socktest.pl
-r-r-r- 1 nick users 1874 Jan 19 10:23 socktest.pl
```

Strange Numbers...

You might have encountered things like chmod 755 somefile and of course you will be wondering what this is. The thing is, that you can change the entire permission pattern of a file in one go using one number like the one in this example. Every mode has a corresponding code number, and as we shall see there is a very simple way to figure out what number corresponds to any mode.

Every one of the three digits on the mode number corresponds to one of the three permission triplets. (u, g and o) Every permission bit in a triplet corresponds to a value: 4 for r, 2 for w, 1 for x. If the permission bit you add this value to the number of the permission triplet. If it is cleared, then you add nothing. (Some of you might notice that in fact, the number for a triplet is the *octal* value corresponding to the three-bit pattern - if you don't know what an octal value is, it doesn't really matter, just follow the intstructions) So if a file has rwxr-xr-x permissions we do the following calculation:

Triplet for u: rwx = > 4 + 2 + 1 = **7**

Triplet for g: r-x = > 4 + 0 + 1 = **5**

Tripler for o: r-x = > 4 + 0 + 1 = **5**

Which makes : 755

So, 755 is a terse way to say *'I don't mind if other people read or run this file, but only I should be able to modify it'* and 777 means *'everyone has full access to this file'*

3.12 PROFILES AND LOCKING ACCOUNTS

Linux Locking an Account

The syntax is as follows:

```
passwd -l {username}
```

The -l option disables an account by changing the password to a value which matches no possible encrypted value. In this example, lock user account named vivek. First, login as a root user and type the following command:

```
# passwd -l vivek
```

Linux Unlocking an Account

The syntax is as follows:

```
passwd -u {username}
```

The -u option re-enables an account by changing the password back to its previous value i.e. to value before using -l option. To unlock user account named vivek. Login as a root user and type following command:

```
# passwd -u vivek
```

3.13 SETTING PASSWORDS

In order to communicate with the UNIX command shell, you first need to learn as to how to log in to your account. As you already know, UNIX is a multi-user system; you can start by finding a terminal, a computer or any other way to access the system. You first have to go through the login process no matter how you access UNIX. When you actually login, you will use a name and password initially assigned by the system administrator.

Steps to Login

1. Logging in is a procedure that tells the UNIX System who you are; the system responds by asking you the password. So, in order to log in, first, connect your PC to the UNIX system. After a successful connection is established, you would find the following prompt coming on the screen.

   Login: Each user on the UNIX system is assigned an account name which identifies him as a unique user. The account name has eight characters or less and is usually based on the first name or the last name. It can have any combination of letters and numbers.
2. Thus, if you want to access the UNIX resources, you should know your account name first. If you don't as yet have an account name, ask the system administrator to assign you one. Now, at the login prompt, enter your account name. Press Enter Key. Type your account name in lowercase letters. UNIX treats uppercase letters differently from lowercase letters.

   ```
   Login : pankaj
   Password:
   ```
3. Once the login name is entered, UNIX prompts you to enter a password. While you are entering your password, it will not be shown on the screen. This is just a security measure adopted by UNIX . The idea behind is that people standing around you are not able to look through the secret password by looking at the screen. Be careful while you are typing your password because you will not be able to see what you have typed. However, if you give either the login name or the password wrong, then UNIX denies you the permission to access its resources. The system then shows an error message on the screen which is given below:

   ```
   Login : pankaj
   Password:
   Login incorrect
   Login:
   ```

   Many UNIX systems give you three or four chances to enter your login and password correct. So, key in your correct login name and the password again.
4. Once you have successfully logged on by giving a correct login and password, you are given some information about the system, some news for users and a message whether you have an electronic mail or not.

   ```
   Login: shefali
   Password:
   Last login: Sun May 14 19:00:29 on ttyAe
   You have mail
   $
   ```

   The dollar sign is the UNIX's method of telling that it's ready to accept commands from the user. You can have a different prompt also in a case where your system is configured for showing a different prompt. By default a $ is shown for the Korn or Bourne Shells.

5. At this point, you are ready to enter your first UNIX command. Now, when you are done working on your UNIX system and decide to leave your terminal-then it is always a good idea to log off the system. It is very dangerous to leave your system without logging out because some mischievous minds could tamper with your files and directories. They can delete your files. They can also read through your private files. Thus, logging off the system is always a better idea than turning off you terminal. In order to log off the system, type the following command:

   ```
   $ exit
   login:
   ```

   The above command will work if you are using a Bourne or a Korn shell. However, if you are working on C shell, exit will work or you can give another command to log off.

   ```
   $ logout
   login:
   ```

3.14 CHANGING YOUR PASSWORD

You can change your password with the 'passwd' command. It is advisable to keep changing your password every few days. This helps in maintaining security. UNIX is very careful about the whole procedure of changing passwords. First UNIX wants you to enter your old password to make sure that a valid user is changing the password. The procedure of changing passwords is very simple. In order to change your password, you first have to log on to the UNIX system. Then issue the 'passwd' command at the UNIX prompt as shown below:

```
$passwd
Changing password for shefali
Enter old password:
Enter new password:
Re-type new password:
$
```

Thus UNIX wants you to type in your old password. Then it asks for the new password. Finally, UNIX confirms your new password by asking you to type in the new password again. If by any means, any mismatch happens, then the UNIX system warns you that the information provided by you is inconsistent as shown below:

```
$ passwd
Changing password for shefali
Enter old password:
Enter new password:
Re-type new password:
Mismatch-password unchanged
$
```

As stated earlier also, UNIX doesn't show the passwords typed by you on the screen because of security reasons.

3.15 CHOOSING A NEW LOGIN NAME AND PASSWORD

It is said to be a good practice to choose good secret passwords or combinations and keep them secure. Here are a few tips for selecting the right login name and password.

1. Your login name should not be less than two characters and normally should not exceed eight characters. Moreover, it should begin with a lowercase.
2. Choose a password that is greater than at-least six characters. The smaller the password, the more vulnerable it is to be leaked out.
3. Don't at all use easily guessable words like sun, moon, stars, etc.
4. Don't choose a password based on personal information like your spouses name, your middle name, your surname, your job title, etc.
5. The password should contain two alphabets and one numeric or a special character. It cannot include spaces in it.
6. Never choose a password that is very complicated and not easily memorizable. In such a case, the chances of leaving a copy of the password on a piece of paper near the computer rise. Therefore, keep a password that is neither too short, nor too difficult to remember.
7. You should avoid picking up a password from the computer dictionary.

You should make sure that you remember your password. When you enter your password, UNIX encrypts it and saves it in a specific file. In a case, if you forget your password, then even the system administrator cannot decrypt it. Therefore, he will have to create an entirely new account for you. Now, you must have realized how important security is on a UNIX system. UNIX also offers a variety of tools to maintain security. One such tool is the usage of the 'lock' command. This command is found on many systems. The lock command locks your keyboard till the time you enter a valid password, as shown below:

```
$lock
Password:
Sorry
Password:
```

If the correct password is not entered, then the lock command does not give control of the keyboard. If the correct password is not entered in a certain number of chances, then the lock command logs you off the system automatically. It is a very powerful tool to use if you are leaving your system just for a few minutes and want to make sure that no mischievous hands fiddle with your files and directories.

3.16 SWITCHING USER, REMOVING USER AND SWITCHING GROUP

Adding New User

First you need to create three (test1,test2,admin1) groups for our examples using groupadd

Options

-d home directory

-s starting program (shell)

-p password

-g (primary group assigned to the users)

-G (Other groups the user belongs to)

-m (Create the user's home directory)

To add a new user with

a primary group of test1

a second group test2

starting shell /bin/bash

password of xxxx

home directory of admin

create home directory

a login name of admin

#useradd -g test1 -G test2 -s /bin/bash -p xxxx -d/home/admin -m admin

This will create a new user admin.

One additional switch worth mentioning is "-D", which controls the defaults for useradd.

Specifying the "-D" switch on its own will simply display the default settings, while specifying -D in conjunction with other switches will change the defaults to those values.

```
# useradd -D
GROUP=100
INACTIVE=-1HOME=/home
INACTIVE=-1
EXPIRE=
SHELL=/bin/bash
SKEL=/etc/skel
# useradd -D -s /bin/sh
# useradd -D
GROUP=100
HOME=/home
INACTIVE=-1
EXPIRE=
SHELL=/bin/sh
SKEL=/etc/skel
```

As you can see, this changes the default shell for created users from "bash" to "sh".

adduser - User Friendly Frontend for useradd command

Syntax

adduser [options] user group

If you want to know available option refer add user man page

Example

```
#adduser admin
Adding user 'admin' ...
Adding new group 'admin' (1001) ...
Adding new user 'admin' (1001) with group 'admin' ...
Creating home directory '/home/admin' ...
Copying files from '/etc/skel' ...
Enter new UNIX password:
Retype new UNIX password:
```

passwd: password updated successfully
Changing the user information for admin
Enter the new value, or press ENTER for the default
Full Name []: Admin
Room Number []:
Work Phone []:
Home Phone []:
Other []:
Is the information correct? [y/N] y

You'll notice that, by default, the adduser command creates a group with the same name as the username, and makes this group the primary group for that user. This is called a user private group (UPG)

Modify User in Linux

usermod - Modify a user account

Syntax

```
usermod [-c comment] [-d home_dir [-m]] [-e expire_date] [-f inactive_days]
[-g initial_group] [-G group [,...]] [-l login_name] [-p passwd] [-s shell] [-u uid [-o]] [-L|-U] login
```

If you want to know available option check usermod man page

Example

Options:
-d home directory
-s starting program (shell)
-p password
-g (primary group assigned to the users)
-G (Other groups the user belongs to)
To add the group 'others' to the user admin
#usermod -G others admin

Delete User in Linux

userdel - Delete a user account and related files

Syntax

userdel [-r] login
If you want to know available options check userdel man page

Example

Options
-r (remove home directory)
To remove the user 'admin' and his home directory
#userdel -r admin

deluser - remove a user from the system

Syntax

deluser [options] user group

If you want more options check deluser man page

Example

By default, deluser will remove the user without removing the home directory, the mail spool or any other files on the system owned by the user. Removing the home directory and mail spool can be achieved using the —remove-home option. If the —home option is given, deluser will only remove the user if the directory given to the —home option matches the user's real home directory.

#deluser —remove-home admin

Groups Administration in Linux

Add New Group in Linux

groupadd - Create a new group

Syntax

groupadd [-g gid [-o]] group

For more options check groupadd man page

Example

#groupadd test1

This will create a test1 group

addgroup - add a group to the system

Syntax

addgroup [options] [—gid ID] group

If you want to know available options check addgroup man page

#addgroup

Enter a groupname to add: admin1Adding group 'admin12†(1001)...Done.

Modify Group in Linux

groupmod - Modify a group

Syntax

groupmod [-g gid [-o]] [-n group_name] group

For more options check groupmod man page

Example

#groupmod test1 test2

This will modify group name test1 to test2

Delete group in Linux

groupdel - Delete a group

Syntax

groupdel groupname

For more options check groupdel man page

Example

#groupdel test2

this will delete the test2 group

delgroup - remove a group from the system

Syntax

delgroup [options] [—only-if-empty] group

For more details about options check delgroup man page

Example

#delgroup --only-if-empty test2

Removing group 'test22 ...done.

groups Command

print the groups a user is in

Syntax

groups [username]

This simple command displays what groups a user is a member of. It takes the username of user as a parameter. If no username is given, it defaults to the current user.

groups

root

groups admin

test1 : test2

SUMMARY

1. The user information is stored in the system /etc/passwd and /etc/shadow files, and that additionally, group membership information is stored in the /etc/group file.
2. The update-passwd tool keeps the entries in these master files in sync on all Debian systems.While it is possible to edit these files by hand, it is not recommended.There exist several command line tools, which can be used to manage these files instead.
3. When a new person joins an organisation he is usually given an account by the system administrator. This is the login account of the user.
4. Now a days almost all Unix systems support an admin tool which seeks the following information from the system administrator to open a new account:
   1. Username: This serves as the login name for the user.
   2. Password: Usually a system administrator gives a simple password.
5. The users are advised to later select a password which they feel comfortable using. User's password appears in the shadow files in encrypted forms.
6. If the file "/etc/nologin" exists and the user is not root, the contents of the "/etc/nologin" file are printed to the screen and the login is terminated.
7. If special access restrictions are specified for the user logging in in the file "etc/usertty", the restrictions must be met or the log in will be denied and the program syslog will log the attempt. If the user is root the login must be on a terminal listed in the file "etc/securetty".
8. The login program will use the account field to find the username and therefore get the UID of the user. Login will also use the password (or the /etc/shadow file) to be sure the entered password is a match.
9. Login will look up the user's home directory and use that to set the $HOME environment variable. Login will use the shell field to determine what shell program (such as bash, sh, tsh, etc) to run for that user. then login will pass program control to the shell program.
10. There is an important difference in the control passed at this point, however! The shell program will run with the user's privileges and not with root privileges.

SHORT QUESTIONS WITH ANSWER

1. Use *the* jobs command to verify whether you have any processes running in background.

Ans. job (maybe the catfun is still running?)

2. Use vi to create a little text file. Suspend vt in background.

Ans.
```
vi text.txt
(inside vi press ctrl-z)
```

3. Verify with Jobs that vi is suspended in background.

Ans.
```
[paul@rhel53 -]$ jobs
[1]+ Stopped        vim text.txt
```

4. Start find / > allfiles.txt 2>/dev/null in foreground. Suspend it in background before it finishes.

Ans.
```
[paul@rhel53-]$ find / > allfiles.txt 2>/dev/null
(press ctrl-z)
[?]+ Stopped            find / > allfiles.txt 2> /dev/null
```

5. Start two long sleep processes in background.

Ans. sleep 4000 & /sleep 5000 &

6. Display all Jobs in background.

Ans.
```
[paul@rhel53 ~]$ jobs
[1]-    Stopped        vim text.txt
[2]+    Stopped        find / > allfiles.txt 2> /dev/null
[3]     Running        sleep 4000 &
[4]     Running        sleep 5000 &
```

7. Use the kill command to suspend the last sleep process.

Ans.
```
[paul@rhel53 ~]$ kill -SIGSTOP 4519
[paul@rhelS3 ~]$ jobs
[1] Stopped                vim text.txt
[2]- Stopped               find />allfiles.txt 2> /dev/null
[3] Running                sleep 4000 &
[4]+ Stopped               sleep 5000
```

8. Continue the find process in background (make sure it runs again),

Ans. bg 2 (verify the Job-id in your jobs list)

9. Put one of the sleep commands back in foreground.

Ans. fg 3 (again verity your job-id)

10. (if time permits, a general review question...) Explain in detail where the numbers come from in the next screenshot. When are the variables replaced by their value ? By which shell ?

Ans.
```
[paul@RHEL4b ~]$ echo $$ $PPID
4224 4223
[paul@RHEL4b ~]$ bash -c "echo $$ $PPID"
4224 4223
[paul@RHEL4b ~]$ bash -c 'echo $$ SPPID'
5059 4224
[paul@RHEL4b ~]$ bash -c 'echo $$ SPPID'
4223: 4224: command not found
```

The current bash shell will replace the $$ and SPPID while scanning the line, and before executing the echo command.

```
[paul@RHEL4b ~]$ echo $$ $PPID
4224 4223
```

The variables are now double quoted, but the current bash shell will replace $$ and $PPID while scanning the line, and before executing the bach -c command.

```
[paul@RHEL4b ~]$ bash -c "echo $$ $PPID"
4224 4223
```

The variables are now single quoted. The current bash shell will **not** replace the $$ and the $PPID. The bash -c command will be executed before the variables replaced with their value. This latter bash is the one replacing the $$ and $PPID with their value.

```
[paul@RHEL4b ~]$ bash -c 'echo $$ $PPID'
5059 4224
```

With backticks the shell will still replace both variable before the embedded echo is executed. The result of this echo is the two process id's. These are given as commands to bash -c. But two numbers are not commands!

```
[paul«RHEL4b ~]$ bash -c `echo $$ $PPID`
4223: 4224: command not found
```

EXERCISE

1. Where is all data stored which is given in by user?
2. What information is stored in an Inode?
3. What do you understand by Indoes?
4. How many types of File owner are there in Unix?
5. How may files can a user make?
6. Define Unix file structure.
7. What is the difference between a Directory & a file in Unix?
8. What are various steps involved in logging into UNIX
9. Can we get to know about last login in UNIX.
10. Can we view the Password in UNIX.
11. Give the command to log off in UNIX.

❐❐❐

CHAPTER 4

MANAGING UNIX FILE SYSTEMS

4.0 INTRODUCTION

Linux requires you to create one or more **partitions**. The next paragraphs will explain how to create and use partitions.

A partition's **geometry** and size is usually defined by a starting and ending cylinder (sometimes by sector). Partitions can be of type **primary** (maximum four), **extended** (maximum one) or **logical** (contained within the extended partition). Each partition has a **type field** that contains a code. This determines the computers operating system or the partitions file system.

4.1 PARTITIONS

Data is commonly stored on magnetic or optical **disk platters**. The platters are rotated (at high speeds). Data is read by **heads**, which are very close to the surface of the platter, without touching it! The heads are mounted on an arm (sometimes called a comb or a fork).

Data is written in concentric circles called **tracks**. Track zero is (usually) on the outside. The time it takes to position the head over a certain track is called the **seek time**. Often the platters are stacked on top of each other, hence the set of tracks accessible at a certain position of the comb forms a **cylinder**. Tracks are divided into 512 byte **sectors**, with more unused space (**gap**) between the sectors on the outside of the platter.

When you break down the advertised **access time** of a hard drive, you will notice that most of that time is taken by movement of the heads (about 65%) and **rotational latency** (about 30%).

Chapter Outline

4.2 BLOCK DEVICE

Random access hard disk devices have an abstraction layer called **block device** to enable formatting in fixed-size (usually 512 bytes) blocks. Blocks can be accessed independent of access to other blocks. A block device has the letter b to denote the file type in the output of **ls -l**.

```
[root@RHEL4b ~]# ls -l /dev/sda*
brw-rw-- 1 root disk 8, 0 Aug 4 22:55 /dev/sda
brw-rw-- 1 root disk 8, 1 Aug 4 22:55 /dev/sda1
brw-rw-- 1 root disk 8, 2 Aug 4 22:55 /dev/sda2
[root@RHEL4b ~]#
```

Note that a **character device** is a constant stream of characters, being denoted by a c in **ls -l**.

Note also that the **ISO 9660** standard for cdrom uses a **2048 byte** block size. Old hard disks (and floppy disks) use **cylinder-head-sector** addressing to access a sector on the disk. Most current disks use **LBA (Logical Block Addressing)**.

ide or scsi

Actually, the title should be **ata** or **scsi**, since ide is an ata compatible device. Most desktops use **ata devices**, most servers use **scsi**.

ata

An **ata controller** allows two devices per bus, one **master** and one **slave**. Unless your controller and devices support **cable select**, you have to set this manually with jumpers.

With the introduction of **sata** (serial ata), the original ata was renamed to **parallel ata**. Optical drives often use **atapi**, which is an ATA interface using the SCSI communication protocol.

scsi

A **scsi controller** allows more than two devices. When using **SCSI (small computer system interface)**, each device gets a unique **scsi id**. The **scsi controller** also needs a **scsi id**, do not use this id for a scsi-attached device.

Older 8-bit SCSI is now called **narrow**, whereas 16-bit is **wide**. When the bus speeds was doubled to 10 Mhz, this was known as **fast SCSI**. Doubling to 20 Mhz made it **ultra SCSI**.

4.3 PRIMARY, EXTENDED AND LOGICAL

Linux requires you to create one or more **partitions**. The next paragraphs will explain how to create and use partitions.

A partition's **geometry** and size is usually defined by a starting and ending cylinder (sometimes by sector). Partitions can be of type **primary** (maximum four), **extended** (maximum one) or **logical** (contained within the extended partition). Each partition has a **type field** that contains a code. This determines the computers operating system or the partitions file system.

TABLE 4.1. Primary, extended and logical partitions

Partition	Type naming
Primary (max 4)	1-4
Extended (max 1)	1-4
Logical	5-

4.4 PARTITION NAMING

We saw before that hard disk devices are named /dev/hdx or /dev/sdx with x depending on the hardware configuration. Next is the partition number, starting the count at 1. Hence the four (possible) primary partitions are numbered 1 to 4. Logical partition counting always starts at 5. Thus /dev/hda2 is the second partition on the first ATA hard disk device, and /dev/hdb5 is the first logical partition on the second ATA hard disk device. Same for SCSI, /dev/sdb3 is the third partition on the second SCSI disk.

TABLE 4.2. Partition naming

partition device
/dev/hda1 first primary partition on /dev/hda
/dev/hda2 second primary or extended partition on /dev/hda
/dev/sda5 first logical drive on /dev/sda
/dev/sdb6 second logical on /dev/sdb

4.5 DISCOVERING PARTITIONS

fdisk -l

In the **fdisk -l** example below you can see that two partitions exist on **/dev/sdb**. The first partition spans 31 cylinders and contains a Linux swap partition. The second partition is much bigger.

```
root@laika:~# fdisk -l /dev/sdb
Disk /dev/sdb: 100.0 GB, 100030242816 bytes
255 heads, 63 sectors/track, 12161 cylinders
Units = cylinders of 16065 * 512 = 8225280 bytes
Device Boot Start End Blocks Id System
/dev/sdb1 1 31 248976 82 Linux swap / Solaris
/dev/sdb2 32 12161 97434225 83 Linux
root@laika:~#
```

/proc/partitions

The **/proc/partitions** file contains a table with major and minor number of partitioned devices, their number of blocks and the device name in **/dev**. Verify with **/proc/ devices** to link the major number to the proper device.

```
paul@RHELv4u4:~$ cat /proc/partitions
major minor #blocks name
3     0     524288   hda
3     64    734003   hdb
8     0     8388608  sda
```

```
8    1   104391   sda1
8    2   8281507  sda2
8    16  1048576  sdb
8    32  1048576  sdc
8    48  1048576  sdd
253  0   7176192  dm-0
253  1   1048576  dm-1
```

The **major** number corresponds to the device type (or driver) and can be found in **/proc/devices**. In this case 3 corresponds to **ide** and 8 to **sd**. The **major** number determines the **device driver** to be used with this device.

The **minor** number is a unique identification of an instance of this device type. The **devices.txt** file in the kernel tree contains a full list of major and minor numbers.

other tools

You might be interested in alternatives to fdisk like **parted**, **cfdisk**, **sfdisk** and **gparted**. This course mainly uses **fdisk** to partition hard disks.

4.6 PARTITIONING NEW DISKS

In the example below, we bought a new disk for our system. After the new hardware is properly attached, you can use **fdisk** and **parted** to create the necessary partition(s).

This example uses **fdisk**, but there is nothing wrong with using **parted**.

recognising the disk

First, we check with **fdisk -l** whether Linux can see the new disk. Yes it does, the new disk is seen as /dev/sdb, but it does not have any partitions yet.

```
root@RHELv4u2:~# fdisk -l
Disk /dev/sda: 12.8 GB, 12884901888 bytes
255 heads, 63 sectors/track, 1566 cylinders
Units = cylinders of 16065 * 512 = 8225280 bytes
Device Boot Start End Blocks Id System
/dev/sda1 * 1 13 104391 83 Linux
/dev/sda2 14 1566 12474472+ 8e Linux LVM
Disk /dev/sdb: 1073 MB, 1073741824 bytes
255 heads, 63 sectors/track, 130 cylinders
Units = cylinders of 16065 * 512 = 8225280 bytes
Disk /dev/sdb doesn't contain a valid partition table
```

opening the disk with fdisk

Then we create a partition with fdisk on /dev/sdb. First we start the fdisk tool with /dev/sdb as argument. Be very very careful not to partition the wrong disk!!

```
root@RHELv4u2:~# fdisk /dev/sdb
Device contains neither a valid DOS partition table, nor Sun, SGI...
Building a new DOS disklabel. Changes will remain in memory only,
until you decide to write them. After that, of course, the previous
content won't be recoverable.
Warning: invalid flag 0x0000 of partition table 4 will be corrected...
```

empty partition table

Inside the fdisk tool, we can issue the **p** command to see the current disks partition table.

```
Command (m for help): p
Disk /dev/sdb: 1073 MB, 1073741824 bytes
255 heads, 63 sectors/track, 130 cylinders
Units = cylinders of 16065 * 512 = 8225280 bytes
Device Boot Start End Blocks Id System
```

create a new partition

No partitions exist yet, so we issue **n** to create a new partition. We choose p for primary, 1 for the partition number, 1 for the start cylinder and 14 for the end cylinder.

```
Command (m for help): n
Command action
e extended
p primary partition (1-4)
p
Partition number (1-4): 1
First cylinder (1-130, default 1):
Using default value 1
Last cylinder or +size or +sizeM or +sizeK (1-130, default 130): 14
```

We can now issue p again to verify our changes, but they are not yet written to disk. This means we can still cancel this operation! But it looks good, so we use **w** to write the changes to disk, and then quit the fdisk tool.

```
Command (m for help): p
Disk /dev/sdb: 1073 MB, 1073741824 bytes
255 heads, 63 sectors/track, 130 cylinders
Units = cylinders of 16065 * 512 = 8225280 bytes
Device Boot Start End Blocks Id System
/dev/sdb1 1 14 112423+ 83 Linux
Command (m for help): w
The partition table has been altered!
Calling ioctl() to re-read partition table.
Syncing disks.
root@RHELv4u2:~#
```

display the new partition

Let's verify again with **fdisk -l** to make sure reality fits our dreams. Indeed, the screenshot below now shows a partition on /dev/sdb.

```
root@RHELv4u2:~# fdisk -l
Disk /dev/sda: 12.8 GB, 12884901888 bytes
255 heads, 63 sectors/track, 1566 cylinders
Units = cylinders of 16065 * 512 = 8225280 bytes
Device Boot Start End Blocks Id System
/dev/sda1 * 1 13 104391 83 Linux
/dev/sda2 14 1566 12474472+ 8e Linux LVM
```

```
Disk /dev/sdb: 1073 MB, 1073741824 bytes
255 heads, 63 sectors/track, 130 cylinders
Units = cylinders of 16065 * 512 = 8225280 bytes
Device Boot Start End Blocks Id System
/dev/sdb1 1 14 112423+ 83 Linux
root@RHELv4u2:~#
```

about the partition table

master boot record

The **partition table** information (primary and extended partitions) is written in the **master boot record** or **mbr**. You can use **dd** to copy the mbr to a file.

This example copies the master boot record from the first SCSI hard disk.

```
dd if=/dev/sda of=/SCSIdisk.mbr bs=512 count=1
```

The same tool can also be used to wipe out all information about partitions on a disk.

This example writes zeroes over the master boot record.

```
dd if=/dev/zero of=/dev/sda bs=512 count=1
```

Or to wipe out the whole partition or disk.

```
dd if=/dev/zero of=/dev/sda
```

partprobe

Don't forget that after restoring a **master boot record** with **dd**, that you need to force the kernel to reread the partition table with **partprobe**. After running **partprobe**, the partitions can be used again.

```
[root@RHEL5 ~]# partprobe
[root@RHEL5 ~]#
```

logical drives

The **partition table** does not contain information about **logical drives**. So the **dd** backup of the **mbr** only works for primary and extended partitions. To backup the partition table including the logical drives, you can use **sfdisk**.

This example shows how to backup all partition and logical drive information to a file.

```
sfdisk -d /dev/sda > parttable.sda.sfdisk
```

The following example copies the **mbr** and all **logical drive** info from /dev/sda to /dev/sdb.

```
sfdisk -d /dev/sda | sfdisk /dev/sdb
```

4.7 SWAP SPACE

Creating a swap space

A swap file is an ordinary file; it is in no way special to the kernel. The only thing that matters to the kernel is that it has no holes, and that it is prepared for use with **mkswap**. It must reside on a local disk, however; it can't reside in a filesystem that has been mounted over NFS due to implementation reasons.

The bit about holes is important. The swap file reserves the disk space so that the kernel can quickly swap out a page without having to go through all the things that are necessary when allocating a disk sector to a file. The kernel merely uses any sectors that have already been allocated to the file. Because a hole in a file means that there are no disk sectors allocated (for that place in the file), it is not good for the kernel to try to use them.

One good way to create the swap file without holes is through the following command:

```
$ dd if=/dev/zero of=/extra-swap bs=1024
count=1024
1024+0 records in
1024+0 records out
$
```

where /extra-swap is the name of the swap file and the size of is given after the count=. It is best for the size to be a multiple of 4, because the kernel writes out *memory pages*, which are 4 kilobytes in size. If the size is not a multiple of 4, the last couple of kilobytes may be unused.

A swap partition is also not special in any way. You create it just like any other partition; the only difference is that it is used as a raw partition, that is, it will not contain any filesystem at all. It is a good idea to mark swap partitions as type 82 (Linux swap); this will the make partition listings clearer, even though it is not strictly necessary to the kernel.

After you have created a swap file or a swap partition, you need to write a signature to its beginning; this contains some administrative information and is used by the kernel. The command to do this is **mkswap**, used like this:

```
$ mkswap /extra-swap 1024
Setting up swapspace, size = 1044480
bytes
$
```

Note that the swap space is still not in use yet: it exists, but the kernel does not use it to provide virtual memory.

You should be very careful when using **mkswap**, since it does not check that the file or partition isn't used for anything else. *You can easily overwrite important files and partitions with **mkswap**!* Fortunately, you should only need to use **mkswap** when you install your system.

The Linux memory manager limits the size of each swap space to 2 GB. You can, however, use up to 8 swap spaces simultaneously, for a total of 16 GB

4.7.1 Using a Swap Space

An initialized swap space is taken into use with **swapon**. This command tells the kernel that the swap space can be used. The path to the swap space is given as the argument, so to start swapping on a temporary swap file one might use the following command.

```
$ swapon /extra-swap
$
```

Swap spaces can be used automatically by listing them in the /etc/fstab file.

```
/dev/hda8     none     swap     sw     0     0
/swapfile     none     swap     sw     0     0
```

The startup scripts will run the command **swapon -a**, which will start swapping on all the swap spaces listed in **/etc/fstab**. Therefore, the **swapon** command is usually used only when extra swap is needed.

You can monitor the use of swap spaces with **free**. It will tell the total amount of swap space used.

```
$ free
          total    used    free    shared
buffers
Mem:      15152    14896   256     12404    2528
-/+ buffers:       12368   2784
Swap:     32452    6684    25768
$
```

The first line of output (Mem:) shows the physical memory. The total column does not show the physical memory used by the kernel, which is usually about a megabyte. The used column shows the amount of memory used (the second line does not count buffers). The free column shows completely unused memory. The shared column shows the amount of memory shared by several processes; the more, the merrier. The buffers column shows the current size of the disk buffer cache.

That last line (Swap:) shows similar information for the swap spaces. If this line is all zeroes, your swap space is not activated.

The same information is available via **top**, or using the proc filesystem in file /proc/meminfo. It is currently difficult to get information on the use of a specific swap space.

A swap space can be removed from use with **swapoff**. It is usually not necessary to do it, except for temporary swap spaces. Any pages in use in the swap space are swapped in first; if there is not sufficient physical memory to hold them, they will then be swapped out (to some other swap space). If there is not enough virtual memory to hold all of the pages Linux will start to thrash; after a long while it should recover, but meanwhile the system is unusable. You should check (e.g., with **free**) that there is enough free memory before removing a swap space from use.

All the swap spaces that are used automatically with **swapon -a** can be removed from use with **swapoff -a**; it looks at the file /etc/fstab to find what to remove. Any manually used swap spaces will remain in use.

Sometimes a lot of swap space can be in use even though there is a lot of free physical memory. This can happen for instance if at one point there is need to swap, but later a big process that occupied much of the physical memory terminates and frees the memory. The swapped-out data is not automatically swapped in until it is needed, so the physical memory may remain free for a long time. There is no need to worry about this, but it can be comforting to know what is happening.

4.8 DEVICE FILES

- Under Linux, tape drives are accessed through several groups of device files
- Each device group has number, with the first drive numbered 0
- These are the most commonly used devices:
  - */dev/st0* — SCSI tape drive, which will be automatically rewound after each operation

- */dev/nst0* — the same drive, but with no automatic rewinding
- */dev/ft0* — floppy tape drive
- */dev/nft0* — the same without rewinding
- */dev/ht0* — ATAPI tape drive
- */dev/nht0* — the same without rewinding

Once the most common Linux file systems is the **ext2** (the second extended) file system. A disadvantage is that file system checks on ext2 can take a long time.

You will see that ext2 is being replaced by **ext3** on most Linux machines. They are essentially the same, except for the **journaling** which is only present in ext3.

Journaling means that changes are first written to a journal on the disk. The journal is flushed regularly, writing the changes in the file system. Journaling keeps the file system in a consistent state, so you don't need a file system check after an unclean shutdown or power failure.

You can create these file systems with the **/sbin/mkfs** or **/sbin/mke2fs** commands. Use **mke2fs -j** to create an ext3 file system. You can convert an ext2 to ext3 with **tune2fs -j**. You can mount an ext3 file system as ext2, but then you lose the journaling. Do not forget to run **mkinitrd** if you are booting from this device.

ext4

Since, 2008, the newest incarnation of the ext file system is **ext4** is available in the Linux kernel. **ext4** support larger files (up to 16 terabyte) and larger file systems than **ext3** (and many more features).

vfat

The **vfat** file system exists in a couple of forms: **fat12** for floppy disks, **fat16** on **msdos**, and **fat32** for larger disks. The Linux **vfat** implementation supports all of these, but vfat lacks a lot of features like security and links. **fat** disks can be read by every operating system, and are used a lot for digital cameras, **usb** sticks and to exchange data between different OS'ses on a home user's computer.

iso 9660

iso 9660 is the standard format for cdroms. Chances are you will encounter this file system also on your hard disk in the form of images of cdroms (often with the .iso extension). The **iso 9660** standard limits filenames to the 8.3 format. The Unix world didn't like this, and thus added the **rock ridge** extensions, which allows for filenames up to 255 characters and Unix-style file-modes, ownership and symbolic links. Another extensions to **iso 9660** is **joliet**, which adds 64 unicode characters to the filename. The **el torito** standard extends **iso 9660** to be able to boot from CDROM's.

udf

Most optical media today (including cd's and dvd's) use **udf**, the Universal Disk Format.

swap

All things considered, swap is not a file system. But to use a partition as a **swap partition** it must be formatted and mounted as swap space.

4.9 RAW AND BLOCK FILES

Blocks and Fragments

The Boot Block

The boot block stores the procedures used in booting the system. If a file system is not to be used for booting, the boot block is left blank. The boot block appears only in the first cylinder group (cylinder group 0) and is the first 8 Kbytes in a partition.

Most of the file system is taken up by data blocks, which contain whatever the users have put in their files.

The hardware disk sector is usually 512 bytes. A block size larger than 512 bytes is desirable for speed. However, because UNIX file system usually contain a very large number of small files, much larger blocks would cause excessive internal fragmentation. That is why the earlier 4.1 BSD file system was limited to 1024 – byte block.

The 4.2 BSD solution is to use two block sizes for files which have no indirect blocks: all the blocks of the file are large block size except the last. The last block is an appropriate multiple of a smaller fragment size to fill out the file. Thus, a file of size 18000 bytes would have two 8K blocks and one 2K block fragment.

The block and fragment sizes are set during the file – creation according to the intended use of the file system: if many small files are expected, the fragment size should be small; if repeated transfer of large files are expected, the basic block size should be large.

4.10 FORMATTING DISKS AND SOME OTHER COMMANDS

You can use the fdisk utility to toggle the bootable flag (change the boot partition). This can help if you install an OS over linux that wipes LILO. However the OS that wiped LILO must be able to toggle this partition to a Linux filesystem.

Programs used to manage filesystems

- badblocks(8) - Search a device for badblocks. The command "badblocks /dev/hda" will search the first partition of the first IDE hard drive for badblocks.
- cfdisk(8) - A partition table manipulator used to create or delete disk partitions.
- dosfsck(8) - Used to check a msdos filesystem.
- dumpe2fs(8) - Lists the superblock and blocks group information on the device listed. Use with a command like "dumpe2fs /dev/hda2". The filesystem on the device must be a Linux filesystem for this to work.
- fdformat(8) - Performs s lowlevel format on a floppy disk. Ex: "fdformat /dev/fd0H1440".
- fdisk(8) - Used to add or remove partitions on a disk device. It modifies the partition table entries.
- fsck(8) - Used to check and/or repair a Linux filesystem. This should only be used on systems that are not mounted.
- hdparm(8) - Used to get or set the hard disk parameters.
- mkdosfs(8) - Used to create a msdos filesystem.
- mke2fs(8) - Create a Linux native filesystem which is called a second extended filesystem. This creates the current version of the Linux filesystem.
- mkfs(8) - Used to make a Linux filesystem on a device. The command "mkfs /dev/hdb1" will create a Linux filesystem on the first partition of the second IDE drive.
- mkswap(8) - Creates a Linux swap area on a device.
- mount(8) - Used to mount a filesystem. It supports many types of filesystems.
- stat(1u) - Used to print out inode information on a file. Usage: stat filename
- swapoff(8) - Used to de-activate a swap partition.
- swapon(8) - Used to activate a swap partition.

- tune2fs(8) - Used to adjust filesystem parameters that are tunable on a Linux second extended filesystem. The filesystem must not be mounted write when this operation is performed. Can adjust maximum mount counts between filesystem checks, the time between filesystem checks, the amount of reserved blocks, and other parameters.
- umount(8) - Unmount a filesystem.

4.11 MAKING FILE SYSTEMS

You need to create UFS file systems only occasionally. The system software automatically creates file systems as part of the installation process. You need to create (or re-create) a UFS file system when you:

- Add or replace disks
- Change the slices of an existing disk
- Do a full restore on a file system
- Change some parameters (such as block size) of a file system

Use the newfs command to create UFS file systems. newfs is a convenient front-end to the mkfs command, the program that creates the new file system.

On Solaris 2.x systems, information used to set some parameter defaults, such as number of tracks per cylinder and number of sectors per track, is read from the disk label. newfs determines the file system parameters to use based on the options you choose and information from the disk label, and passes the parameters to the mkfs command, which builds the file system.

Although you can use the mkfs command directly, you need to know much more about the parameters and how to choose them. mkfs supplies defaults for all parameters, but they are not tuned to the underlying hardware. You must also specify the total size of the slice.

Use the mkfs command:

- To create a disk-based file system type other than UFS, if the software for such a file system type is available.
- To create a UFS file system whose logical geometry differs from the physical geometry of the disk. For example, you could specify parameters appropriate for a diskette.

How to Create a UPS File System

The disk must be formatted and divided into slices before you can create UFS file systems on it. If you are re-creating an existing UFS file system, unmount the file system before performing these steps:

1. Become superuser.
2. Type newfs /dev/rdsk/device-name and press Return.

   You are asked if you want to proceed.

   Caution: Be sure you have specified the correct device name for the slice before performing the next step. If you specify the wrong slice, you will erase its contents when the new file system is created.
3. Type y to confirm.

   The newfs command uses optimized default values to create the file sytem.

   Example : How to Create a UFS File System

   This example creates a file system on /dev/rdsk/c0t3d0s7:

```
jupiter% su
Password:
#newfs /dev/rdsk/c0t3d0s7
newfs: construct a new file system /dev/rdsk/c0td0s7 (y/n)? y
/dev/rdsk/c0t3d0s7: 163944 sectord in 506 cylinders of 9 tracks, 36
sectors
83.9MB in 32 cyl groups (16 c/g, 2.65MB/g, 1216 i/g)
super-block backups (for fsck -b #) at :
32, 5264, 10496, 15728, 20960, 26192, 31424, 36656, 41888,
47120, 52352, 57584, 62816, 68048, 73280, 78512, 82976, 88208,
93440, 98672, 103904, 109136, 114368, 119600, 124832, 130064, 135296,
140528, 145760, 150992, 156224, 161456,
#
```

How to Create an Entry in the File Sytem Table

1. Become superuser.
2. Edit the /etc/vfstab file.
3. Add the entry, separating each field with white space (a space or a Tab). If a field has no contents, enter a dash (-).
4. Save the changes.
5. Check to be sure the mount point directory is present. If not, create it:
   (*a*) Change to the directory where you want to create the mount point.
   (*b*) Type mkdir directory-name and press Return.
6. Type mount mount-point and press Return.

   The entry is mounted.

This example mounts the disk slice /dev/dsk/c0t3d0s7 as a UFS file system attached to the mount point directory /files1 with the default mount options (read/write). It specifies the raw character device /dev/rdsk/c0t3d0s7 as the device to fsck. The fsck pass value of 2 means that the file system will be checked, but not sequentially.

```
# device            device              mount      ps     fsck     auto-     mount
# to mount          to fsck             point      type   pass     mount?    options
/dev/dsk/c0t3d0s7  /dev/rdsk/c0t3d0s7  /files1    ufs 2           yes
```

This example mounts the directory /export/man from the system pluto as an NFS file system on mount point /usr/man. You do not specify a device to fsck or a fsck pass for NFS file systems. In this example, mount options are ro (read-only) and soft. For greater reliability, specify the hard mount for read/write NFS file systems.

```
#device             device      mount          fs     fsck     auto-      mount
#to mount           to fsck     point          type   pass     mount?     options
pluto:/export/man   -           /usr/man       nfs    -        yes        ro, soft
```

This example mounts the root file system on a loopback mount point named /tmp/newroot. Specify yes for automount, no device to fsck, and no fsck pass number. Loopback file systems must always be mounted after the file systems used to make up the loopback file system. Be sure that the loopback entry is the last entry in the /etc/vfstab file so that it follows the entries it depends on.

```
#device      device    mount          fs    fsck   auto-    mount
#to mount    to fsck   point          type  pass   mount?   options
/            -         /tmp/newroot   lofs  -      yes      -
```

You do not need to put entries in the /etc/vfstab file for CD-ROM or floppy disk devices. CD-ROM devices are mounted automatically by the Volume Management software. Diskettes are mounted after you run the volcheck(1) command.

Media Type	Mount Location	State of Media
floppy	/floppy/floppy0	symbolic link to mounted floppy in the first local floppy drive
floppy	/floppy/floppy1	if there are two floppy drives present, symbolic link to mounted floppy in the second local floppy drive
	/floppy/floppy_name	mounted named floppy
	/floppy/unnamed_floppy	mounted unnamed floppy
CD-ROM	/cdrom/cdrom0	symbolic link to mounted CD-ROM in the first local CD-ROM drive
CD ROM	/cdrom/cdrom1	if there is a second CD-ROM drive, symbolic link to mounted CD-ROM in second local CD-ROM drive
	/cdrom/CD-ROM_name	mounted named CD-ROM
	/cdrom/CD-ROM_name/slice	mounted named CD-ROM with a file system
	/cdrom/unnamed_cdrom	mounted unnamed CD-ROM

Media Type	Device Location	State of Media
floppy	/vol/dev/diskette0/unnamed _floppy	formatted unnamed floppy–block device access
	/vol/dev/rdiskette0/unnamed _floppy	formatted unnamed floppy–raw device acess
	/vol/dev/diskette0/unlabeled	unlabeled floppy–block device acess
	/vol/dev/rdiskette0/unlabeled	unlabeled floppy–raw device acess
CD-ROM	/vol/dev/dsk/c0t6/unnamed_cdrom	CD-ROM–block device acess
	/vol/dev/rdsk/c0t6/unnamed_cdrom	CD-ROM–raw device acess

4.12 SUPERBLOCK I-NODES

The Superblock

The superblock stores much of the information about the file system. A few of the more important things it contains are:

- Size and status of the file system
- Label (file system name and volume name)
- Size of the file system logical block
- Date and time of the last update
- Cylinder group size
- Number of data blocks in a cylinder group
- Summary data block

- File system state: clean, stable, or active
- Path name of the last mount point

The superblock is located at the beginning of the disk partition, and is replicated in each cylinder group. Because the superblock contains critical data, multiple superblocks are made when the file system is created. Each of the superblock replicas is offset by a different amount from the beginning of its cylinder group. For multiple-platter disk drives, the offsets are calculated so that a superblock appears on each platter of the drive. That way, if the first platter is lost, an alternate superblock can always be retrieved. Except for the leading blocks in the first cylinder group, the leading blocks created by the offsets are used for data storage.

A summary information block is kept with the superblock. It is not replicated, but is grouped with the first superblock, usually in cylinder group 0. The summary block records changes that take place as the file system is used, and lists the number of inodes, directories, fragments, and storage blocks within the file system.

4.13 INODES

An inode contains all the information about a file except its name, which is kept in a directory. An inode is 128 bytes. The inode information is kept in the cylinder information block, and contains:

- The type of the file
- Regular
- Directory
- Block special
- Character special
- Symbolic link
- FIFO, also known as named pipe
- Socket
- The mode of the file (the set of read-write-execute permissions)
- The number of hard links to the file
- The user-id of the owner of the file
- The group-id to which the file belongs
- The number of bytes in the file
- An array of 15 disk-block addresses
- The date and time the file was last accessed
- The date and time the file was last modified
- The date and time the file was created

The array of 15 disk addresses (0 to 14) point to the data blocks that store the contents of the file. The first 12 are direct addresses; that is, they point directly to the first 12 logical storage blocks of the contents of the file. If the file is larger than 12 logical blocks, the 13th address points to an indirect block, which contains direct block addresses instead of file contents. The 14th address points to a double indirect block, which contains addresses of indirect blocks. The 15th address is for triple indirect addresses, if they are ever needed.

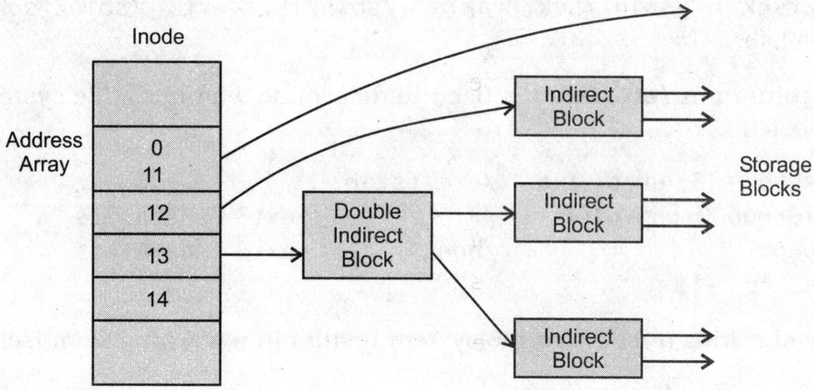

4.14 STORAGE BLOCKS

The rest of the space allocated to the file system is occupied by storage blocks, also called data blocks. The size of these storage blocks is determined at the time a file system is created. Storage blocks are allocated, by default, in two sizes: an 8-Kbyte logical block size, and a 1-Kbyte fragmentation size.

For a regular file, the storage blocks contain the contents of the file. For a directory, the storage blocks contain entries that give the inode number and the file name of the files in the directory.

4.15 FREE BLOCKS

Blocks not currently being used as inodes, as indirect address blocks, or as storage blocks are marked as free in the cylinder group map. This map also keeps track of fragments to prevent fragmentation from degrading disk performance.

To give you an idea of the appearance of a typical UFS file system,

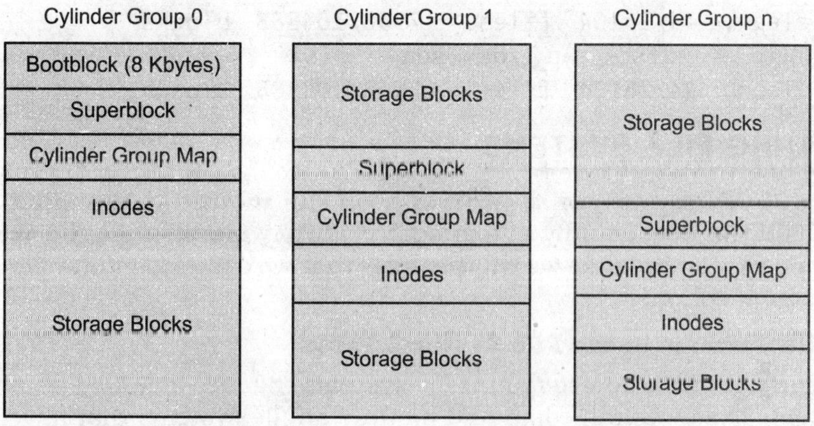

4.16 FILE SYSTEM CHECKER

The **fsck** command is a front end tool used to check a file system for errors.

```
[root@RHEL4b -1# ls /sbin/*fsck*
/sbin/dosfsck      /sbin/fsck      /sbin/fsck.ext2  /sbin/fsck.msdos
```

```
/sbin/e2fsck   /sbin/fsck.cramfs  /sbin/fsck.ext3  /sbin/fsck.vfat
[root@RHEL4b -1#
```

The last column in **/etc/fstab** is used to determine whether a file system should be checked at boot-up.

```
[paul@RHELb -]$ grep ext  /etc/fstab
/dev/VolGroup00/LogVol00   /             ext3  defaults     11
LABEL=/boot                /boot         ext3  defaykts     12
[paul@RHEL4b -]$
```

Manually checking a mounted file system results in a warning from fsck.

```
[root@RHEL4b~]# fsck /boot
fsck 1.35 (28-Feb-2004)
e2fsck 1.35 (28-Feb-2004)
/dev/sdal is mounted.
WARNIN!!! Running e2fsck on a mounted filesystem may cause
SEVERE filesystem damage.
Do you really want to continue (y/n)? no
check aborted.
```

But after unmounting fsck and **e2fsck** can be used an ex2 file sytem.

```
[root@RHEL4b ~]# fsck /boot
fsck 1.35 (28-Feb-2004)
e2fsck 1.35 (28-Feb-2004)
/boot: clean, 44/26104 files, 17598/104388 blocks
[root@RHEL4b~]# fsck -p /boot
fsck 1.35 (28-Feb-2004)
/boot: clean, 44/26104 files, 17598/104388 blocks
[root@RHELb~]# e2fsck -p /dev/sdal
/boot: clean, 44/26104 files, 17598/104388 blocks
```

4.17 MOUNTING FILE SYSTEMS

Once you've put a file system on a partition, you can **mount** it. Mounting a file system makes it available for use, usually as a directory. We say **mounting a file system** instead of mounting a partition because we will see later that we can also mount file systems that do not exists on partitions.

Mounting File Systems in the File Systems Table

This section describes how to mount file systems that are included in the /etc/vfstab file for a system. The mountall command mounts all file systems that have a yes in that automount field. The mountall command is run automatically when entering multiuser run states. Use the mountall command to mount all file systems, all local file systems, all remote file systems, or all file systems of a particular file system type in the /etc/vfstab file.

Note: The mount-point directory must exist before you can mount any file system on it.

To create a mount point, change to the directory where you want to create the mount point, and type mkdir mount-point and press Return.

How to Mount File Systems in the File System Table

1. Become superuser.
2. Type mountall and press Return.

   All the file systems with a fsck device entry are checked and fixed, if necessary, before mounting. All file systems listed in the /etc/vfstab file with yes in the automount field are mounted.

```
mars%
Password:
# mountall
#
```

If file system are already mounted, messages are displayed telling you so.

```
mars% su
Password:
# mountall
mount: /tmp already mounted
nfs mount: mount: /usr/openwin: Device busy
nfs mount: mount: /usr/man: Device busy
#
```

4.18 MOUNTING LOCAL FILE SYSTEMS

On all **Unix** systems, every file and every directory is part of one big file tree. To access a file, you need to know the full path starting from the root directory. When adding a **file system** to your computer, you need to make it available somewhere in the file tree. The directory where you make a file system available is called a **mount point**.

/bin/mkdir

This example shows how to create a new **mount point** with **mkdir**.

```
root@RHELv4u2:~# mkdir /home/project55
```

/bin/mount

When the **mount point** is created, and a **file system** is present on the partition, then **mount** can **mount** the **file system** on the **mount point directory**.

```
root@RHELv4u2:~# mount -t ext2 /dev/sdb1 /home/project55/
```

Once mounted, the new file system is accessible to users.

/etc/filesystems

Actually the explicit **-t ext2** option to set the file system is not always necessary. The **mount** command is able to automatically detect a lot of file systems.

When mounting a file system without specifying explicitly the file system, then **mount** will first probe **/etc/filesystems**. Mount will skip lines with the **nodev** directive.

```
paul@RHELv4u4:~$ cat /etc/filesystems
ext3
ext2
```

```
nodev  proc
nodev  devpts
iso9660
vfat
hfs
paul@RHELv4u4:~$
```

/proc/filesystems

When **/etc/filesystems** does not exist, or ends with a single * on the last line, then **mount** will read **/proc/filesystems**.

```
[root@RHEL52 ~]#  cat /proc/filesystems | grep -v ^nodev
ext2
iso9660
ext3
```

/bin/umount

You can **unmount** a mounted file system using the **umount** command.

```
root@pasha:~# umount /home/reet
```

4.19 DISPLAYING MOUNTED FILE SYSTEMS

To display all mounted file systems, issue the **mount** command. Or look at the files **/proc/mounts** and **/etc/mtab**.

/bin/mount

The simplest and most common way to view all mounts is by issuing the **mount** command without any arguments.

```
root@RHELv4u2:~# mount | grep /dev/sdb
/dev/sdb1 on /home/project55 type ext2 (rw)
```

/proc/mounts

The kernel provides the info in **/proc/mounts** in file form, but **/proc/mounts** does not exist as a file on any hard disk. Looking at **/proc/mounts** is looking at information that comes directly from the kernel.

```
root@RHELv4u2:~# cat /proc/mounts | grep /dev/sdb
/dev/sdb1 /home/project55 ext2 rw 0 0
```

/etc/mtab

The **/etc/mtab** file is not updated by the kernel, but is maintained by the **mount** command. Do not edit **/etc/mtab** manually.

```
root@RHELv4u2:~# cat /etc/mtab | grep /dev/sdb
/dev/sdb1 /home/project55 ext2 rw 0 0
```

/bin/df

A more user friendly way to look at mounted file systems is **df**. The **df (diskfree)** command has the added benefit of showing you the free space on each mounted disk.

Like a lot of Linux commands, **df** supports the **-h** switch to make the output more **human readable**.

```
root@RHELv4u2:~# df
Filesystem 1K-blocks Used Available Use% Mounted on
```

```
/dev/mapper/VolGroup00-LogVol00
11707972 6366996 4746240 58% /
/dev/sda1 101086 9300 86567 10% /boot
none 127988 0 127988 0% /dev/shm
/dev/sdb1 108865 1550 101694 2% /home/project55
root@RHELv4u2:~# df -h
Filesystem Size Used Avail Use% Mounted on
/dev/mapper/VolGroup00-LogVol00
12G 6.1G 4.6G 58% /
/dev/sda1 99M 9.1M 85M 10% /boot
none 125M 0 125M 0% /dev/shm
/dev/sdb1 107M 1.6M 100M 2% /home/project55
```

In the **df -h** example below you can see the size, free space, used gigabytes and percentage and mount point of a partition.

```
root@laika:~# df -h | egrep -e "(sdb2|File)"
Filesystem Size Used Avail Use% Mounted on
/dev/sdb2 92G 83G 8.6G 91% /media/sdb2
root@laika:~#
```

/bin/du

The **du** command can summarize **disk usage** for files and directories. Preventing **du** to go into subdirectories with the **-s** option will give you a total for that directory.

This option is often used together with **-h**, so **du -sh** on a mount point gives the total amount used in that partition.

```
root@pasha:~# du -sh /home/reet
881G /home/reet
```

permanent mounts

Until now, we performed all mounts manually. This works nice, until the next reboot.

Luckily there is a way to tell your computer to automatically mount certain file systems during boot.

/etc/fstab

This is done using the file system table located in the **/etc/fstab** file. Below is a sample **/etc/fstab** file.

```
roor@RHELv4i2:~# cat /etc/fstab
/dev/VolGroup00/LogVol00 /            ext3    defaults          1 1
LABEL=/boot              /boot        ext3    defaults          1 2
none                     /dev/pts     devpts  gid=5,mode=620    0 0
none                     /dev/shm     tmpfs   defaults          0 0
none                     /proc        proc    defaults          0 0
none                     /sys         sysfs   defaults          0 0
/dev/VolGroup0/LogVol01  swap         swap    defaults          0 0
```

By adding the following line, we can automate the mounting of a file system.

```
/dev/sdb1       /home/project55     ext2    defaults        0 0
```

mount /mountpoint

Adding an entry to **/etc/fstab** has the added advantage that you can simplify the **mount** command. The command in the screenshot below forces **mount** to look for the partition info in **/etc/fstab**.

```
# mount /home/project55
```

4.20 LOGICAL VOLUMES

The logical block size is the size of the blocks that the UNIX kernel uses to read or write files. The logical block size is usually different from the physical block size (usually 512 bytes), which is the size of the smallest block that the disk controller can read or write.

You can specify the logical block size of the file system. Once the file system is created, you cannot change this parameter without rebuilding the file system.

You can have file systems with different logical block sizes on the same disk. By default, the logical block size is 8192 bytes (8 Kbytes) for UFS file systems.

The UFS file system supports block sizes of 4096 or 8192 bytes (4 or 8 Kbytes). 8 Kbytes is the recommended logical block size.

To choose the best logical block size for your system, consider both the performance desired and the available space. For most UFS systems, an 8 Kbyte file system provides the best performance, offering a good balance between disk performance and use of space in primary memory and on disk.

As a general rule, to increase efficiency, use a larger logical block size for file systems where most of the files are very large. Use a smaller logical block size for file systems where most of the files are very small. You can use the quot -c *file-system* command on a file system to display a complete report on the distribution of files by block size.

4.21 NETWORK FILE SYSTEMS

The **network file system** (or simply **nfs**) enables us since the eighties to share a directory with other computers on the network.

In this chapter we see how to setup an **nfs** server and and **nfs** client computer.

4.21.1 nfs Protocol Versions

The older **nfs** versions 2 and 3 are stateless (**udp**) by default (but they can use **tcp**).

The more recent **nfs version 4** brings a stateful protocol with better performance and stronger security.

NFS version 4 was defined in **rfc 3010** in 2000 and **rfc 3530** in 2003 and requires tcp (port 2049). It also supports **Kerberos** user authentication as an option when mounting a share. NFS versions 2 and 3 authenticate only the host.

rpcinfo

Clients connect to the server using **rpc** (on Linux this can be managed by the **portmap** daemon. Look at **rpcinfo** to verify that **nfs** and its related services are running.

```
root®RHELv4u2:~# /etc/init.d/portmap status
portmap (pid 1920) is running...
root®RHELv4u2:~# rpcinfo -p
program     vers proto   port
100000      2    tcp     111       portmapper
```

```
100000      2      udp       111       portmapper
100024      1      udp       32768     status
100024      1      tcp       32769     status
root®RHELv4u2:~# service nfs start
Starting NFS services:                              [ OK ]
Starting NFS quotas:                                [ OK ]
Starting NFS daemon:                                [ OK ]
Starting NFS mountd:                                [ OK ]
```

The same **rpcinfo** command when nfs is started.

```
rootffiRHELv4u2 : ~# rpcinfo -p
program  vers  proto    port
100000   2     tcp      111       portmapper
100000   2     udp      111       portmapper
100024   1     udp      32768     status
100024   1     tcp      32769     status
100011   1     udp      985       rquotad
100011   2     udp      985       rquotad
100011   1     tcp      988       rquotad
100011   2     tcp      988       rquotad
100003   2     udp      2049      nfs
100003   3     udp      2049      nfs
100003   4     udp      2049      nfs
100003   2     tcp      2049      nfs
100003   3     tcp      2049      nfs
100003   4     tcp      2049      nfs
100021   1     udp      32770     nlockmgr
100021   3     udp      32770     nlockmgr
100021   4     udp      32770     nlockmgr
100021   1     tcp      32789     nlockmgr
100021   3     tcp      32789     nlockmgr
100021   4     tcp      32789     nlockmgr
100005   1     udp      1004      mountd
100005   1     tcp      1007      mountd
100005   2     udp      1004      mountd
100005   2     tcp      1007      mountd
100005   3     udp      1004      mountd
100005   3     tcp      1007      mountd
root@RHELv4u2: ~#
```

server configuration

nfs is configured in **/etc/exports**. You might want some way (**ldap**?) to synchronize userid's across computers when using **nfs** a lot.

The **rootsquash** option will change UID 0 to the UID of a **nobody** (or similar) user account. The **sync** option will write writes to disk before completing the client request.

/etc/exports

Here is a sample **/etc/exports** to explain the syntax:

```
paul@laika:~$ cat /etc/exports
# Everyone can read this share
/mnt/data/iso *(ro)
# Only the computers named pasha and barry can readwrite this one
/var/www pasha(rw) barry(rw)
# same, but without root squashing for barry
/var/ftp pasha(rw) barry(rw,no_root_squash)
# everyone from the netsec.local domain gets access
/var/backup *.netsec.local(rw)
# ro for one network, rw for the other
/var/upload 192.168.1.0/24(ro) 192.168.5.0/24(rw)
```

More recent incarnations of **nfs** require the **subtree_check** option to be explicitly set (or unset with **no_subtree_check)**. The **/etc/exports** file then looks like this:

```
root@debian6 ~# cat /etc/exports
# Everyone can read this share
/srv/iso *(ro,no_subtree_check)
# Only the computers named pasha and barry can readwrite this one
/var/www pasha(rw,no_subtree_check) barry(rw,no_subtree_check)
# same, but without root squashing for barry
/var/ftp pasha(rw,no_subtree_check) barry(rw,no_root_squash,no_subtree_
check)
```

exportfs

You don't need to restart the nfs server to start exporting your newly created exports.

You can use the **exportfs -va** command to do this. It will write the exported directories to **/var/lib/nfs/etab**, where they are immediately applied.

```
root@debian6 ~# exportfs -va
exporting pasha:/var/ftp
exporting barry:/var/ftp
exporting pasha:/var/www
exporting barry:/var/www
exporting *:/srv/iso
```

4.22 BOOT DISKS

Boot disk contains the booting program, the booting process using boot program is known as Booting the System:

Booting is the process of starting up a computer. It is so called because the job involves something akin to lifting oneself up by one's bootstraps. That is, from a system, which is off you, need to have a system running a complex operating system like UNIX. This task is done in stages, with a very simple program in the computer's ROM, which runs and loads a program on the hard disk, which is then able to load the whole of the UNIX Kernel.

Of course the UNIX Kernel and other utilities cannot reach the hard disk by them selves. A system administrator would do well to know how to set up his system from the scratch. On a small system this is usually possible from floppies, while on a large computer this task could be done from a CD – ROM or any higher capacity storage medium.

What happens when you first start up the computer?

The boot program is usually located in the block zero of the boot device. This program is loaded and run by a program in the ROM of the computer and gives a message akin to the following:

Enter Kernel to Boot

Here the boot program is asking you which file on the disk contains the UNIX kernel program you want to run. Usually this file is called UNIX and is located in the root device. But you can have other filename, which you want to use to start up. Typically this is done because you have configured a somewhat different kernel tuned to be useful in different situations. Then you can specify that filename. Sometimes the kernel is assumed to be in UNIX and you have to take special action if you want to boot from a different kernel.

After telling which kernel is to be loaded, the machine starts running that kernel and first brings you to a system state known as the single user state. In this state only one user can be working on the machine and only one console is active. No multi user capability will exist at this time. The person at the console will be super user and will have all the rights and powers of one. This mode is also called system maintenance mode because it is only in this state that some actions can be performed. It is usually in this state that the system administrator sets the system date. To set the date

 # date 980607104533

where the first two are the year, the next two are the month, the next two are the days in the month, followed by the time in hours, minute, and seconds. There are some variations to the format in which the date can be set.

You are now ready to enter the multi user mode. The transition from mode to another is achieved by running a program called init. This provides for 7 run level ranging from 0 to 6 and an s or S state for the single user mode. The main multi user mode is level 2, and so when the system start and asks you

 # Enter new run level

you should enter 2 to reach normal multi user mode.

Run level 0 is the down state and readies the machine for powering off. You can use the mode with the init command at the # prompt to switch off the computer. The who r option tells you the current run level of your system.

The devices and users available in each run level are controlled by a file called /etc/inittab and it also specifies what action is to be taken when a run level is entered.

Once you reach run level 2, all the computers are connected to the server will display the usual login prompt. It is now possible for users to login and work on these terminals.

SUMMARY

1. Linux requires you to create one or more **partitions**. The next paragraphs will explain how to create and use partitions.
2. A partition's **geometry** and size is usually defined by a starting and ending cylinder (sometimes by sector). Partitions can be of type **primary** (maximum four), **extended** (maximum one) or **logical** (contained within the extended partition).

3. Each partition has a **type field** that contains a code. This determines the computers operating system or the partitions file system.
4. Data is commonly stored on magnetic or optical **disk platters**. The platters are rotated (at high speeds).
5. Data is read by **heads**, which are very close to the surface of the platter, without touching it! The heads are mounted on an arm (sometimes called a comb or a fork).
6. Data is written in concentric circles called **tracks**. Track zero is (usually) on the outside. The time it takes to position the head over a certain track is called the **seek time**. Often the platters are stacked on top of each other, hence the set of tracks accessible at a certain position of the comb forms a **cylinder**.
7. Tracks are divided into 512 byte **sectors**, with more unused space (**gap**) between the sectors on the outside of the platter.
8. When you break down the advertised **access time** of a hard drive, you will notice that most of that time is taken by movement of the heads (about 65%) and **rotational latency** (about 30%).
9. An **ata controller** allows two devices per bus, one **master** and one **slave**. Unless your controller and devices support **cable select**, you have to set this manually with jumpers.
10. With the introduction of **sata** (serial ata), the original ata was renamed to **parallel ata**. Optical drives often use **atapi**, which is an ATA interface using the SCSI communication protocol.

SHORT QUESTIONS WITH ANSWERS

1. Use dmesg to make a list of hard disk devices detected at boot-up.

Ans. Some possible answers...

```
dmesg | grep -i disk
Looking for ATA disks: dmesg |  grep hd[abcd]
Looking for ATA disks: dmesg |  grep -i "ata disk"
Looking for SCSI disks: dmesg | grep sd[a-f]
Looking for SCSI disks: dmesg | grep -i "scsi disk"
```

2. Use fdisk to find the total size of all hard disk devices on your system.

Ans. `fdisk -l`

3. Stop a virtual machine, add three virtual 1 gigabyte scsi hard disk devices and one virtual 400 megabyte ide hard disk device. If possible, also add another virtual 400 megabyte ide disk.

Ans. This exercise happens in the settings of vmware or VirtualBox.

4. Use dmesg to verify that all the new disks are properly detected at boot-up.

Ans. See 1.

5. Verify that you can see the disk devices in /dev.

Ans.

```
SCSI+SATA: ls -l /dev/sd*
ATA: ls -l /dev/hd*
```

6. Use fdisk (with grep and /(dev/null) to display the total size of the new disks.

Ans.

```
root@rhel53 ~# fdisk -l 2>/dev/null | grep [MGT]B
Disk /dev/hda: 21.4  GB,   21474836480 bytes
Disk /dev/hdb: 1073  MB,   1073741824 bytes
Disk /dev/sda: 2147  MB,   2147483648 bytes
Disk /dev/sdb: 2147  MB,   2147483648 bytes
Disk /dev/sdc: 2147  MB,   2147483648 bytes
```

7. Use badblocks to completely erase one of the smaller hard disks.

Ans.

```
#Verify the device (/dev/sdc??) you want to erase before typing this.
```

```
#
root@rhel53 ~# badblocks -ws /dev/sdc
Testing with pattern Oxaa: done
Reading and comparing: done
Testing with pattern 0x55: done
Reading and comparing: done
Testing with pattern Oxff: done
Reading and comparing: done
Testing with pattern 0x00: done
Reading and comparing: done
```

8. Look at /proc/scsi/scsi.

Ans.

```
root@rhel53 ~# cat /proc/scsi/scsi
Attached devices:
Host: scsi0 Channel: 00 Id: 02 Lun: 00
  Vendor: VBOX Model: HARDDISK                    Rev: 1.0
  Type: Direct-Access                             ANSI SCSI revision:
Host: scsiO Channel: 00 Id: 03 Lun: 00
  Vendor: VBOX Model: HARDDISK                    Rev: 1.0
  Type: Direct-Access                             ANSI SCSI revision:
Host: scsiO Channel: 00 Id: 06 Lun: 00
  Vendor: VBOX Model: HARDDISK                    Rev: 1.0
  Type: Direct-Access                             ANSI SCSI revision:
```

9. If possible, install lsscsi, lshw and use them to list the disks.

Ans.

```
Debian.Ubuntu: aptitude install lsscsi Ishw
Fedora: yum install lsscsi lshw
root@rhel53 -# Isscsi
[0:0:2:0]    disk    VBOX    HARDDISK    1.0    /dev/sda
[0:0:3:0]    disk    VBOX    HARDDISK    1.0    /dev/sdb
[0:0:6:0]    disk    VBOX    HARDDISK    1.0    /dev/sdc
```

EXERCISE

1. Is unix case sensitive
2. What message appears when wrong password is given.
3. Can you change Passwords in UNIX.
4. Give procedure to change Password in UNIX.
5. What Parameters should be kept in mind while choosing username & passwords?
6. Can any other user change your password.
7. What is Utility of finger command?
8. How many options can be used with WHO command?
9. What is the purpose of password command?
10. How differently col command be used.

❒❒❒

CHAPTER 5

LINUX KERNEL CONFIGURATION

Chapter Outline

5.0 INTRODUCTION

The hardware of a UNIX system is present in the centre of the diagram. It provides the basic services such as memory management, processor execution level, etc. to the operating system. The UNIX system seems to be organized as a set of layers. The system Kernel is also, called the operating system. The user programs are independent of the hardware on which they are running. Therefore, it becomes very easy to run programs on UNIX system running on different hardware if the programs do make special assumptions.

5.1 ABOUT THE LINUX KERNEL VERSIONS

In 1991 Linux Torvalds wrote (the first version of) the Linux kernel. He put it online, and other people started contributing code. Over 4000 individuals contributed source code to the latest kernel release (version 2.6.27 in November 2008).

Major Linux kernel versions used to come in even and odd numbers. Versions 2.0, **2.2, 2.4 and 2.6 are considered stable kernel versions. Whereas 2.1, 2.3 and 2.5 were** unstable (read development) versions. Since, the release of 2.6.0 in January 2004, all development has been done in the 2.6 tree. There is currently no v2.7.x and according to Linus the even/stable vs odd/development scheme is abandoned forever.

uname -r

To see your current Linux kernel version, issue the uname -r command as shown below.

This first example shows Linux major version 2.6 and minor version 24. The rest -22-**generic is specific to the distribution (Ubuntu in this case).**

```
paul@laika:~$ uname -r
2.6.24-22-generic
```

The same command on Red Hat Enterprise Linux shows an older kernel (2.6.18) with -**92.1.17.el5 being specific to the distribution.**

```
[paul@RHEL52 ~]$ uname -r
2.6.18-92.1.17.el5
```

/proc/cmdline

The parameters that were passed to the kernel at boot time are in /proc/cmdline.

```
paul@RHELv4u4:~$ cat /proc/cmdline
ro root=/dev/VolGroup00/LogVol00 rhgb quiet
```

single user mode

When booting the kernel with the single parameter, it starts in single user mode. Linux can start in a bash shell with the root user logged on (without password). Some distributions prevent the use of this feature (at kernel compile time).

init=/bin/bash

Normally the kernel invokes init as the first daemon process. Adding init=/bin/bash to the kernel parameters will instead invoke bash (again with root logged on without providing a password).

/var/log/messages

The kernel reports during boot to syslog which writes a lot of kernel actions in /var/**log/messages. Looking at this file reveals when the kernel was started, including all** the devices that were detected at boot time.

```
[root@RHEL53 ~]# grep -A16 "syslogd 1.4.1:" /var/log/messages|cut -b24-
syslogd 1.4.1: restart.
kernel: klogd 1.4.1, log source = /proc/kmsg started.
kernel: Linux version 2.6.18-128.el5 (mockbuild@hs20-bc1-5.build.red...
kernel: BIOS-provided physical RAM map:
kernel: BIOS-e820: 0000000000000000 - 000000000009f800 (usable)
kernel: BIOS-e820: 000000000009f800 - 00000000000a0000 (reserved)
kernel: BIOS-e820: 00000000000ca000 - 00000000000cc000 (reserved)
kernel: BIOS-e820: 00000000000dc000 - 0000000000100000 (reserved)
kernel: BIOS-e820: 0000000000100000 - 000000001fef0000 (usable)
kernel: BIOS-e820: 000000001fef0000 - 000000001feff000 (ACPI data)
kernel: BIOS-e820: 000000001feff000 - 000000001ff00000 (ACPI NVS)
kernel: BIOS-e820: 000000001ff00000 - 0000000020000000 (usable)
kernel: BIOS-e820: 00000000fec00000 - 00000000fec10000 (reserved)
kernel: BIOS-e820: 00000000fee00000 - 00000000fee01000 (reserved)
kernel: BIOS-e820: 00000000fffe0000 - 0000000100000000 (reserved)
kernel: 0MB HIGHMEM available.
kernel: 512MB LOWMEM available.
```

This example shows how to use /var/log/messages to see kernel information about **/dev/sda.**

```
[root@RHEL53 ~]# grep sda /var/log/messages | cut -b24-
```

```
kernel: SCSI device sda: 41943040 512-byte hdwr sectors (21475 MB)
kernel: sda: Write Protect is off
kernel: sda: cache data unavailable
kernel: sda: assuming drive cache: write through
kernel: SCSI device sda: 41943040 512-byte hdwr sectors (21475 MB)
kernel: sda: Write Protect is off
kernel: sda: cache data unavailable
kernel: sda: assuming drive cache: write through
kernel: sda: sda1 sda2
kernel: sd 0:0:0:0: Attached scsi disk sda
kernel: EXT3 FS on sda1, internal journal
```

dmesg

The dmesg command prints out all the kernel bootup messages (from the last boot).

```
[root@RHEL53 ~]# dmesg | head
Linux version 2.6.18-128.el5 (mockbuild@hs20-bc1-5.build.redhat.com)
BIOS-provided physical RAM map:
 BIOS-e820: 0000000000000000 - 000000000009f800 (usable)
 BIOS-e820: 000000000009f800 - 00000000000a0000 (reserved)
 BIOS-e820: 00000000000ca000 - 00000000000cc000 (reserved)
 BIOS-e820: 00000000000dc000 - 0000000000100000 (reserved)
 BIOS-e820: 0000000000100000 - 000000001fef0000 (usable)
 BIOS-e820: 000000001fef0000 - 000000001feff000 (ACPI data)
 BIOS-e820: 000000001feff000 - 000000001ff00000 (ACPI NVS)
 BIOS-e820: 000000001ff00000 - 0000000020000000 (usable)
```

Thus to find information about /dev/sda, using dmesg will yield only kernel messages from the last boot.

```
[root@RHEL53 ~]# dmesg | grep sda
SCSI device sda: 41943040 512-byte hdwr sectors (21475 MB)
sda: Write Protect is off
sda: Mode Sense: 5d 00 00 00
sda: cache data unavailable
sda: assuming drive cache: write through
SCSI device sda: 41943040 512-byte hdwr sectors (21475 MB)
sda: Write Protect is off
sda: Mode Sense: 5d 00 00 00
sda: cache data unavailable
sda: assuming drive cache: write through
 sda: sda1 sda2
sd 0:0:0:0: Attached scsi disk sda
EXT3 FS on sda1, internal journal
```

5.2 LINUX KERNEL SOURCE

ftp.kernel.org

The home of the Linux kernel source is ftp.kernel.org. It contains all official releases of the Linux kernel source code from 1991. It provides free downloads over http, ftp and rsync of all these releases, as well as changelogs and patches. More information can be otained on the website www.kernel.org.

Anyone can anonymously use an ftp client to access ftp.kernel.org

```
paul@laika:~$ ftp ftp.kernel.org
Connected to pub3.kernel.org.
220 Welcome to ftp.kernel.org.
Name (ftp.kernel.org:paul): anonymous
331 Please specify the password.
Password:
230- Welcome to the
230-
230- LINUX KERNEL ARCHIVES
230- ftp.kernel.org
```

All the Linux kernel versions are located in the pub/linux/kernel/ directory.

```
ftp> ls pub/linux/kernel/v*
200 PORT command successful. Consider using PASV.
150 Here comes the directory listing.
drwxrwsr-x 2 536 536 4096 Mar 20 2003 v1.0
drwxrwsr-x 2 536 536 20480 Mar 20 2003 v1.1
drwxrwsr-x 2 536 536 8192 Mar 20 2003 v1.2
drwxrwsr-x 2 536 536 40960 Mar 20 2003 v1.3
drwxrwsr-x 3 536 536 16384 Feb 08 2004 v2.0
drwxrwsr-x 2 536 536 53248 Mar 20 2003 v2.1
drwxrwsr-x 3 536 536 12288 Mar 24 2004 v2.2
drwxrwsr-x 2 536 536 24576 Mar 20 2003 v2.3
drwxrwsr-x 5 536 536 28672 Dec 02 08:14 v2.4
drwxrwsr-x 4 536 536 32768 Jul 14 2003 v2.5
drwxrwsr-x 7 536 536 110592 Dec 05 22:36 v2.6
226 Directory send OK.
ftp>
```

/usr/src

On your local computer, the kernel source is located in /usr/src. Note though that the structure inside /usr/src might be different depending on the distribution that you are using.

First let's take a look at /usr/src on Debian. There appear to be two versions of the complete Linux source code there. Looking for a specific file (e1000_main.c) with find reveals it's exact location.

```
paul@barry:~$ ls -l /usr/src/
drwxr-xr-x 20 root root 4096 2006-04-04 22:12 linux-source-2.6.15
drwxr-xr-x 19 root root 4096 2006-07-15 17:32 linux-source-2.6.16
paul@barry:~$ find /usr/src -name e1000_main.c
/usr/src/linux-source-2.6.15/drivers/net/e1000/e1000_main.c
/usr/src/linux-source-2.6.16/drivers/net/e1000/e1000_main.c
```

This is very similar to /usr/src on Ubuntu, except there is only one kernel here (and it is newer).

```
paul@laika:~$ ls -l /usr/src/
drwxr-xr-x 23 root root 4096 2008-11-24 23:28 linux-source-2.6.24
paul@laika:~$ find /usr/src -name "e1000_main.c"
/usr/src/linux-source-2.6.24/drivers/net/e1000/e1000_main.c
```

Now take a look at /usr/src on Red Hat Enterprise Linux.

```
[paul@RHEL52 ~]$ ls -l /usr/src/
drwxr-xr-x 5 root root 4096 Dec 5 19:23 kernels
drwxr-xr-x 7 root root 4096 Oct 11 13:22 redhat
```

We will have to dig a little deeper to find the kernel source on Red Hat!

```
[paul@RHEL52 ~]$ cd /usr/src/redhat/BUILD/
[paul@RHEL52 BUILD]$ find . -name "e1000_main.c"
./kernel-2.6.18/linux-2.6.18.i686/drivers/net/e1000/e1000_main.c
```

downloading the kernel source

Debian

Installing the kernel source on Debian is really simple with aptitude install linux-**source. You can do a search for all linux-source packeges first, like in this screenshot.**

```
root@barry:~# aptitude search linux-source
v linux-source -
v linux-source-2.6 -
id linux-source-2.6.15 - Linux kernel source for version 2.6.15
i linux-source-2.6.16 - Linux kernel source for version 2.6.16
p linux-source-2.6.18 - Linux kernel source for version 2.6.18
p linux-source-2.6.24 - Linux kernel source for version 2.6.24
```

And then use aptitude install to download and install the Debian Linux kernel source code.

```
root@barry:~# aptitude install linux-source-2.6.24
```

When the aptitude is finished, you will see a new file named /usr/src/linux-source-**<version>.tar.bz2**

```
root@barry:/usr/src# ls -lh
drwxr-xr-x 20 root root 4.0K 2006-04-04 22:12 linux-source-2.6.15
drwxr-xr-x 19 root root 4.0K 2006-07-15 17:32 linux-source-2.6.16
-rw-r-r- 1 root root 45M 2008-12-02 10:56 linux-source-2.6.24.tar.bz2
```

Ubuntu

Ubuntu is based on Debian and also uses aptitude, so the task is very similar.

```
root@laika:~# aptitude search linux-source
i linux-source - Linux kernel source with Ubuntu patches
v linux-source-2.6 -
i A linux-source-2.6.24 - Linux kernel source for version 2.6.24
root@laika:~# aptitude install linux-source
```

And when aptitude finishes, we end up with a /usr/src/linux-source- **<version>.tar.bz file.**

```
oot@laika:~# ll /usr/src
total 45M
```

```
-rw-r-r- 1 root root 45M 2008-11-24 23:30 linux-source-2.6.24.tar.bz2
```

Red Hat Enterprise Linux

The Red Hat kernel source is located on the fourth source cdrom. The file is called **kernel-2.6.9-42.EL.src.rpm (example for RHELv4u4). It is also available online** at ftp://ftp.redhat.com/pub/redhat/linux/enterprise/5Server/en/os/SRPMS/ (example for RHEL5).

To download the kernel source on RHEL, use this long wget command (on one line, without the trailing \).

```
wget ftp://ftp.redhat.com/pub/redhat/linux/enterprise/5Server/en/os/\
SRPMS/kernel-'uname -r'.src.rpm
```

When the wget download is finished, you end up with a 60M .rpm file.

```
[root@RHEL52 src]# ll
total 60M
-rw-r-r- 1 root root 60M Dec 5 20:54 kernel-2.6.18-92.1.17.el5.src.rpm
drwxr-xr-x 5 root root 4.0K Dec 5 19:23 kernels
drwxr-xr-x 7 root root 4.0K Oct 11 13:22 redhat
```

We will need to perform some more steps before this can be used as kernel source code.

First, we issue the rpm -i kernel-2.6.9-42.EL.src.rpm command to install this Red Hat package.

```
[root@RHEL52 src]# ll
total 60M
-rw-r-r- 1 root root 60M Dec 5 20:54 kernel-2.6.18-92.1.17.el5.src.rpm
drwxr-xr-x 5 root root 4.0K Dec 5 19:23 kernels
drwxr-xr-x 7 root root 4.0K Oct 11 13:22 redhat
[root@RHEL52 src]# rpm -i kernel-2.6.18-92.1.17.el5.src.rpm
```

The we move to the SPECS directory and perform an rpmbuild.

```
[root@RHEL52 ~]# cd /usr/src/redhat/SPECS
[root@RHEL52 SPECS]# rpmbuild -bp -vv -target=i686 kernel-2.6.spec
```

The rpmbuild command put the RHEL Linux kernel source code in /usr/src/redhat/**BUILD/kernel-<version>/.**

```
[root@RHEL52 kernel-2.6.18]# pwd
/usr/src/redhat/BUILD/kernel-2.6.18
[root@RHEL52 kernel-2.6.18]# ll
total 20K
drwxr-xr-x 2 root root 4.0K Dec 6 2007 config
-rw-r-r- 1 root root 3.1K Dec 5 20:58 Config.mk
drwxr-xr-x 20 root root 4.0K Dec 5 20:58 linux-2.6.18.i686
drwxr-xr-x 19 root root 4.0K Sep 20 2006 vanilla
drwxr-xr-x 8 root root 4.0K Dec 6 2007 xen
```

5.3 KERNEL BOOT FILES

vmlinuz

The vmlinuz file in /boot is the compressed kernel.

```
paul@barry:~$ ls -lh /boot | grep vmlinuz
```

```
-rw-r-r-  1  root  root  1.2M  2006-03-06  16:22  vmlinuz-2.6.15-1-486
-rw-r-r-  1  root  root  1.1M  2006-03-06  16:30  vmlinuz-2.6.15-1-686
-rw-r-r-  1  root  root  1.3M  2008-02-11  00:00  vmlinuz-2.6.18-6-686
paul@barry:~$
```

initrd

The kernel uses initrd (an initial RAM disk) at boot time. The initrd is mounted before the kernel loads, and can contain additional drivers and modules. It is a compressed **cpio archive, so you can look at the contents in this way.**

```
root@RHELv4u4:/boot#  mkdir  /mnt/initrd
root@RHELv4u4:/boot#  cp  initrd-2.6.9-42.0.3.EL.img  TMPinitrd.gz
root@RHELv4u4:/boot#  gunzip  TMPinitrd.gz
root@RHELv4u4:/boot#  file  TMPinitrd
TMPinitrd:  ASCII  cpio  archive  (SVR4  with  no  CRC)
root@RHELv4u4:/boot#  cd  /mnt/initrd/
root@RHELv4u4:/mnt/initrd#  cpio  -i  |  /boot/TMPinitrd
4985  blocks
root@RHELv4u4:/mnt/initrd#  ls  -l
total  76
drwxr-xr-x  2  root  root  4096  Feb  5  08:36  bin
drwxr-xr-x  2  root  root  4096  Feb  5  08:36  dev
drwxr-xr-x  4  root  root  4096  Feb  5  08:36  etc
-rwxr-xr-x  1  root  root  1607  Feb  5  08:36  init
drwxr-xr-x  2  root  root  4096  Feb  5  08:36  lib
drwxr-xr-x  2  root  root  4096  Feb  5  08:36  loopfs
drwxr-xr-x  2  root  root  4096  Feb  5  08:36  proc
lrwxrwxrwx  1  root  root  3  Feb  5  08:36  sbin  ->  bin
drwxr-xr-x  2  root  root  4096  Feb  5  08:36  sys
drwxr-xr-x  2  root  root  4096  Feb  5  08:36  sysroot
root@RHELv4u4:/mnt/initrd#
```

System.map

The System.map contains the symbol table and changes with every kernel compile. The symbol table is also present in /proc/kallsyms (pre 2.6 kernels name this file /proc/ksyms).

```
root@RHELv4u4:/boot#  head  System.map-'uname  -r'
00000400  A  __kernel_vsyscall
0000041a  A  SYSENTER_RETURN_OFFSET
00000420  A  __kernel_sigreturn
00000440  A  __kernel_rt_sigreturn
c0100000  A  _text
c0100000  T  startup_32
c01000c6  t  checkCPUtype
c0100147  t  is486
c010014e  t  is386
c010019f  t  L6
root@RHELv4u4:/boot#  head  /proc/kallsyms
```

```
c0100228 t _stext
c0100228 t calibrate_delay_direct
c0100228 t stext
c0100337 t calibrate_delay
c01004db t rest_init
c0100580 t do_pre_smp_initcalls
c0100585 t run_init_process
c01005ac t init
c0100789 t early_param_test
c01007ad t early_setup_test
root@RHELv4u4:/boot#
```

.config

The last file copied to the /boot directory is the kernel configuration used for compilation. This file is not necessary in the /boot directory, but it is common practice to put a copy there. It allows you to recompile a kernel, starting from the same configuration as an existing working one.

5.4 LINUX KERNEL MODULES

about kernel modules

The Linux kernel is a monolithic kernel with loadable modules. These modules contain parts of the kernel used typically for device drivers, file systems and network protocols. Most of the time the necessary kernel modules are loaded automatically and dynamically without administrator interaction.

/lib/modules

The modules are stored in the /lib/modules/<kernel-version> directory. There is a separate directory for each kernel that was compiled for your system.

```
paul@laika:~$ ll /lib/modules/
total 12K
drwxr-xr-x 7 root root 4.0K 2008-11-10 14:32 2.6.24-16-generic
drwxr-xr-x 8 root root 4.0K 2008-12-06 15:39 2.6.24-21-generic
drwxr-xr-x 8 root root 4.0K 2008-12-05 12:58 2.6.24-22-generic
```

<module>.ko

The file containing the modules usually ends in .ko. This screenshot shows the location of the isdn module files.

```
paul@laika:~$ find /lib/modules -name isdn.ko
/lib/modules/2.6.24-21-generic/kernel/drivers/isdn/i4l/isdn.ko
/lib/modules/2.6.24-22-generic/kernel/drivers/isdn/i4l/isdn.ko
/lib/modules/2.6.24-16-generic/kernel/drivers/isdn/i4l/isdn.ko
```

lsmod

To see a list of currently loaded modules, use lsmod. You see the name of each loaded module, the size, the use count, and the names of other modules using this one.

```
[root@RHEL52 ~]# lsmod | head -5
Module Size Used by
autofs4 24517 2
```

```
hidp 23105 2
rfcomm 42457 0
l2cap 29505 10 hidp,rfcomm
```

/proc/modules

/proc/modules lists all modules loaded by the kernel. The output would be too long to display here, so lets grep for the vm module.

We see that vmmon and vmnet are both loaded. You can display the same information with lsmod. Actually lsmod only reads and reformats the output of /proc/modules.

```
paul@laika:~$ cat /proc/modules | grep vm
vmnet 36896 13 - Live 0xffffffff88b21000 (P)
vmmon 194540 0 - Live 0xffffffff88af0000 (P)
paul@laika:~$ lsmod | grep vm
vmnet 36896 13
vmmon 194540 0
paul@laika:~$
```

module dependencies

Some modules depend on others. In the following example, you can see that the nfsd module is used by exportfs, lockd and sunrpc.

```
paul@laika:~$ cat /proc/modules | grep nfsd
nfsd 267432 17 - Live 0xffffffff88a40000
exportfs 7808 1 nfsd, Live 0xffffffff88a3d000
lockd 73520 3 nfs,nfsd, Live 0xffffffff88a2a000
sunrpc 185032 12 nfs,nfsd,lockd, Live 0xffffffff889fb000
paul@laika:~$ lsmod | grep nfsd
nfsd 267432 17
exportfs 7808 1 nfsd
lockd 73520 3 nfs,nfsd
sunrpc 185032 12 nfs,nfsd,lockd
paul@laika:~$
```

insmod

Kernel modules can be manually loaded with the insmod command. This is a very simple (and obsolete) way of loading modules. The screenshot shows insmod loading the fat module (for fat file system support).

```
root@barry:/lib/modules/2.6.17-2-686# pwd
/lib/modules/2.6.17-2-686
root@barry:/lib/modules/2.6.17-2-686# lsmod | grep fat
root@barry:/lib/modules/2.6.17-2-686# insmod kernel/fs/fat/fat.ko
root@barry:/lib/modules/2.6.17-2-686# lsmod | grep fat
fat 46588 0
```

insmod is not detecting dependencies, so it fails to load the isdn module (because the isdn module depends on the slhc module).

```
[root@RHEL52 drivers]# pwd
/lib/modules/2.6.18-92.1.18.el5/kernel/drivers
[root@RHEL52 kernel]# insmod isdn/i4l/isdn.ko
```

```
insmod: error inserting 'isdn/i4l/isdn.ko': -1 Unknown symbol in module
```

modinfo

As you can see in the screenshot of modinfo below, the isdn module depends in the slhc module.

```
[root@RHEL52 drivers]# modinfo isdn/i4l/isdn.ko | head -6
filename: isdn/i4l/isdn.ko
license: GPL
author: Fritz Elfert
description: ISDN4Linux: link layer
srcversion: 99650346E708173496F6739
depends: slhc
```

modprobe

The big advantage of modprobe over insmod is that modprobe will load all necessary modules, whereas insmod requires manual loading of dependencies. Another advantage is that you don't need to point to the filename with full path.

This screenshot shows how modprobe loads the isdn module, automatically loading slhc in background.

```
[root@RHEL52 kernel]# lsmod | grep isdn
[root@RHEL52 kernel]# modprobe isdn
[root@RHEL52 kernel]# lsmod | grep isdn
isdn 122433 0
slhc 10561 1 isdn
[root@RHEL52 kernel]#
```

/lib/modules/<kernel>/modules.dep

Module dependencies are stored in modules.dep.

```
[root@RHEL52 2.6.18-92.1.18.el5]# pwd
/lib/modules/2.6.18-92.1.18.el5
[root@RHEL52 2.6.18-92.1.18.el5]# head -3 modules.dep
/lib/modules/2.6.18-92.1.18.el5/kernel/drivers/net/tokenring/3c359.ko:
/lib/modules/2.6.18-92.1.18.el5/kernel/drivers/net/pcmcia/3c574_cs.ko:
/lib/modules/2.6.18-92.1.18.el5/kernel/drivers/net/pcmcia/3c589_cs.ko:
```

depmod

The modules.dep file can be updated (recreated) with the depmod command. In this screenshot no modules were added, so depmod generates the same file.

```
root@barry:/lib/modules/2.6.17-2-686# ls -l modules.dep
-rw-r-r- 1 root root 310676 2008-03-01 16:32 modules.dep
root@barry:/lib/modules/2.6.17-2-686# depmod
root@barry:/lib/modules/2.6.17-2-686# ls -l modules.dep
-rw-r-r- 1 root root 310676 2008-12-07 13:54 modules.dep
```

rmmod

Similar to insmod, the rmmod command is rarely used anymore.

```
[root@RHELv4u3 ~]# modprobe isdn
[root@RHELv4u3 ~]# rmmod slhc
```

```
ERROR: Module slhc is in use by isdn
[root@RHELv4u3 ~]# rmmod isdn
[root@RHELv4u3 ~]# rmmod slhc
[root@RHELv4u3 ~]# lsmod | grep isdn
[root@RHELv4u3 ~]#
```

modprobe -r

Contrary to rmmod, modprobe will automatically remove unneeded modules.

```
[root@RHELv4u3 ~]# modprobe isdn
[root@RHELv4u3 ~]# lsmod | grep isdn
isdn 133537 0
slhc 7233 1 isdn
[root@RHELv4u3 ~]# modprobe -r isdn
[root@RHELv4u3 ~]# lsmod | grep isdn
[root@RHELv4u3 ~]# lsmod | grep slhc
[root@RHELv4u3 ~]#
```

/etc/modprobe.conf

The /etc/modprobe.conf file and the /etc/modprobe.d directory can contain aliases (used by humans) and options (for dependent modules) for modprobe.

```
[root@RHEL52 ~]# cat /etc/modprobe.conf
alias scsi_hostadapter mptbase
alias scsi_hostadapter1 mptspi
alias scsi_hostadapter2 ata_piix
alias eth0 pcnet32
alias eth2 pcnet32
alias eth1 pcnet32
```

5.5 COMPILING A KERNEL

extraversion

Enter into /usr/src/redhat/BUILD/kernel-2.6.9/linux-2.6.9/ and change the **extraversion in the Makefile.**

```
[root@RHEL52 linux-2.6.18.i686]# pwd
/usr/src/redhat/BUILD/kernel-2.6.18/linux-2.6.18.i686
[root@RHEL52 linux-2.6.18.i686]# vi Makefile
[root@RHEL52 linux-2.6.18.i686]# head -4 Makefile
VERSION = 2
PATCHLEVEL = 6
SUBLEVEL = 18
EXTRAVERSION = -paul2008
```

make mrproper

Now clean up the source from any previous installs with make mrproper. If this is your first after downloading the source code, then this is not needed.

```
[root@RHEL52 linux-2.6.18.i686]# make mrproper
  CLEAN scripts/basic
  CLEAN scripts/kconfig
```

```
CLEAN  include/config
CLEAN  .config  .config.old
```

.config

Now copy a working .config from /boot to our kernel directory. This file contains the configuration that was used for your current working kernel. It determines whether modules are included in compilation or not.

```
[root@RHEL52 linux-2.6.18.i686]# cp /boot/config-2.6.18-92.1.18.el5
.config
```

make menuconfig

Now run make menuconfig (or the graphical make xconfig). This tool allows you to select whether to compile stuff as a module (m), as part of the kernel (*), or not at all (smaller kernel size). If you remove too much, your kernel will not work. The configuration will be stored in the hidden .config file.

```
[root@RHEL52 linux-2.6.18.i686]# make menuconfig
```

make clean

Issue a make clean to prepare the kernel for compile. make clean will remove most generated files, but keeps your kernel configuration. Running a make mrproper at this point would destroy the .config file that you built with make menuconfig.

```
[root@RHEL52 linux-2.6.18.i686]# make clean
```

make bzImage

And then run make bzImage, sit back and relax while the kernel compiles. You can use time make bzImage to know how long it takes to compile, so next time you can go for a short walk.

```
[root@RHEL52 linux-2.6.18.i686]# time make bzImage
  HOSTCC  scripts/basic/fixdep
  HOSTCC  scripts/basic/docproc
  HOSTCC  scripts/kconfig/conf.o
  HOSTCC  scripts/kconfig/kxgettext.o
...
```

This command will end with telling you the location of the bzImage file (and with time info if you also specified the time command.

```
Kernel: arch/i386/boot/bzImage is ready (#1)
real  13m59.573s
user  1m22.631s
sys  11m51.034s
[root@RHEL52 linux-2.6.18.i686]#
```

You can already copy this image to /boot with cp arch/i386/boot/bzImage /boot/ **vmlinuz-<kernel-version>.**

make modules

Now run make modules. It can take 20 to 50 minutes to compile all the modules.

```
[root@RHEL52 linux-2.6.18.i686]# time make modules
  CHK  include/linux/version.h
```

```
CHK  include/linux/utsrelease.h
CC  [M]  arch/i386/kernel/msr.o
CC  [M]  arch/i386/kernel/cpuid.o
CC  [M]  arch/i386/kernel/microcode.o
```

make modules_install

To copy all the compiled modules to /lib/modules just run make modules_install (takes about 20 seconds). Here's a screenshot from before the command.

```
[root@RHEL52 linux-2.6.18.i686]# ls -l /lib/modules/
total 20
drwxr-xr-x 6 root root 4096 Oct 15 13:09 2.6.18-92.1.13.el5
drwxr-xr-x 6 root root 4096 Nov 11 08:51 2.6.18-92.1.17.el5
drwxr-xr-x 6 root root 4096 Dec  6 07:11 2.6.18-92.1.18.el5
[root@RHEL52 linux-2.6.18.i686]# make modules_install
```

And here is the same directory after. Notice that make modules_install created a new directory for the new kernel.

```
[root@RHEL52 linux-2.6.18.i686]# ls -l /lib/modules/
total 24
drwxr-xr-x 6 root root 4096 Oct 15 13:09 2.6.18-92.1.13.el5
drwxr-xr-x 6 root root 4096 Nov 11 08:51 2.6.18-92.1.17.el5
drwxr-xr-x 6 root root 4096 Dec  6 07:11 2.6.18-92.1.18.el5
drwxr-xr-x 3 root root 4096 Dec  6 08:50 2.6.18-paul2008
```

/boot

We still need to copy the kernel, the System.map and our configuration file to /boot. Strictly speaking the .config file is not obligatory, but it might help you in future compilations of the kernel.

```
[root@RHEL52 ]# pwd
/usr/src/redhat/BUILD/kernel-2.6.18/linux-2.6.18.i686
[root@RHEL52 ]# cp System.map /boot/System.map-2.6.18-paul2008
[root@RHEL52 ]# cp .config /boot/config-2.6.18-paul2008
[root@RHEL52 ]# cp arch/i386/boot/bzImage /boot/vmlinuz-2.6.18-paul2008
```

mkinitrd

The kernel often uses an initrd file at bootup. We can use mkinitrd to generate this file. Make sure you use the correct kernel name!

```
[root@RHEL52 ]# pwd
/usr/src/redhat/BUILD/kernel-2.6.18/linux-2.6.18.i686
[root@RHEL52 ]# mkinitrd /boot/initrd-2.6.18-paul2008 2.6.18-paul2008
```

bootloader

Compilation is now finished, don't forget to create an additional stanza in grub or lilo.

5.6 COMPILING ONE MODULE

hello.c

A little C program that will be our module.

```
[root@rhel4a kernel_module]# cat hello.c
#include <linux/module.h>
```

```
#include <section>

int init_module(void)
{
 printk(KERN_INFO "Start Hello World...\n");
 return 0;
}


void cleanup_module(void)
{
 printk(KERN_INFO "End Hello World... \n");
}
```

Makefile

The make file for this module.

```
[root@rhel4a kernel_module]# cat Makefile
obj-m += hello.o
all:
make -C /lib/modules/$(shell uname -r)/build M=$(PWD) modules
clean:
make -C /lib/modules/$(shell uname -r)/build M=$(PWD) clean
```

These are the only two files needed.

```
[root@rhel4a kernel_module]# ll
total 16
-rw-rw-r- 1 paul paul 250 Feb 15 19:14 hello.c
-rw-rw-r- 1 paul paul 153 Feb 15 19:15 Makefile
```

make

The running of the make command.

```
[root@rhel4a kernel_module]# make
make -C /lib/modules/2.6.9-paul-2/build M=~/kernel_module modules
make[1]: Entering dir... '/usr/src/redhat/BUILD/kernel-2.6.9/linux-
2.6.9'
CC [M] /home/paul/kernel_module/hello.o
Building modules, stage 2.
MODPOST
CC /home/paul/kernel_module/hello.mod.o
LD [M] /home/paul/kernel_module/hello.ko
make[1]: Leaving dir... '/usr/src/redhat/BUILD/kernel-2.6.9/linux-
2.6.9'
[root@rhel4a kernel_module]#
```

Now we have more files.

```
[root@rhel4a kernel_module]# ll
total 172
-rw-rw-r- 1 paul paul 250 Feb 15 19:14 hello.c
```

```
-rw-r-r-  1  root  root  64475  Feb  15  19:15  hello.ko
-rw-r-r-  1  root  root  632  Feb  15  19:15  hello.mod.c
-rw-r-r-  1  root  root  37036  Feb  15  19:15  hello.mod.o
-rw-r-r-  1  root  root  28396  Feb  15  19:15  hello.o
-rw-rw-r-  1  paul  paul  153  Feb  15  19:15  Makefile
[root@rhel4a  kernel_module]#
```

hello.ko

Use modinfo to verify that it is really a module.

```
[root@rhel4a  kernel_module]#  modinfo  hello.ko
filename:  hello.ko
vermagic:  2.6.9-paul-2  SMP  686  REGPARM  4KSTACKS  gcc-3.4
depends:
[root@rhel4a  kernel_module]#
```

Good, so now we can load our hello module.

```
[root@rhel4a  kernel_module]#  lsmod  |  grep  hello
[root@rhel4a  kernel_module]#  insmod  ./hello.ko
[root@rhel4a  kernel_module]#  lsmod  |  grep  hello
hello  5504  0
[root@rhel4a  kernel_module]#  tail  -1  /var/log/messages
Feb  15  19:16:07  rhel4a  kernel:  Start  Hello  World...
[root@rhel4a  kernel_module]#  rmmod  hello
[root@rhel4a  kernel_module]#
```

Finally /var/log/messages has a little surprise.

```
[root@rhel4a  kernel_module]#  tail  -2  /var/log/messages
Feb  15  19:16:07  rhel4a  kernel:  Start  Hello  World...
Feb  15  19:16:35  rhel4a  kernel:  End  Hello  World...
[root@rhel4a  kernel_module]#
```

SUMMARY

1. The text segment contains the compiled object instructions, the data segment contains static variables, and the stack segment holds the runtime stack used to store temporary variables.
2. A set of source file that is compiled and linked–into an executable form is stored in a file with the dafault name of a. out (of course, the file can be expl'icitly given any name by the programmer). If the program references statically define data, such as C static variables, a template for the data segment is maintained in the executable file.
3. The data segment will be created and initialized to contain values and space for variables when the executable file is loaded and executed. The stack segment is used to allocate storage for dynamic elements of the program, such as automatic C variables that are created when they come into scope and are destroyed when pass out of scope.
4. The executable file is created by the compiler and linker. These utilities do not define a process; they define only the program text and a template for the data component that the process will use when it executes the program. When the loader loads a program into the computer's memory, the system creates appropriate data and stack segments, called a process.
5. A process has a unique process identifier, a PID, that is essentially a pointer–an integer into a table of process descriptors used by the UNIX OS kernel to reference the process's descriptor.

Whenever one process references another process in a system call, it provides the PID of the target process.

6. The UNIX pa command lists each process associated with the user executing the command. The PID of each process appears as a field in the descriptor of each process. The next time you are using UNIX, try the ps - aux command to observe the PID value identifying each process in the system.
7. The UNIX command for creating a new process is the fork system call. Whenever a (parent) process calls fork, a child process is created with its descriptor, including its own copies of the parent's program text, data, and segments, and access to all open file descriptors (in the kernel).
8. The child and percent processes execute in their own separate address spaces. This means that even though they have access to the same information, both the child and its parent each reference their own copy of the data.
9. No part of the address space of either process is shared. Hence, the parent and child cannot communicate by referencing variables stored at the same address in their respective address space.
10. In UNIX, the only thing the two processes can reference in common is a file.

SHORT QUESTIONS WITH ANSWERS

1. Use fdisk -1 to display existing partitions and sizes.

Ans. `as root: # fdisk -1`

2. Use df-h to display existing partitions and sizes.

Ans. `df -h`

3. Compare the output of fdisk and df.

Ans. Some partitions will be listed in both outputs (maybe /dev/sdal or /dev/hdal).

4. Create a 200MB primary partition on a small disk.

Ans. Choose one of the disks you added (this example uses /dev/sdc).

```
root@rhel53 ~# fdisk /dev/sdc
...
Command (m fox: help) : n
Command action
  e extended
  p primary partition (1-4)
P
Partition number (1-4): 1
First cylinder (1-261, default 1): 1
Last cylinder or +size or +sizeM or +sizeK (1-261, default 261) : +200m
Command (m for help): w
The partition table has been altered!
Calling ioctl() to re-read partition table.
Syncing disks.
```

5. Create a 400MB primary partition and two 300MB logical drives on a big disk.

Ans. Choose one of the disks you added (this example uses /dev/sdb)

```
fdisk /dev/sdb
inside fdisk : n p 1 +400m enter ---n e 2 enter enter ---n 1 +300m (twice)
```

6. Use df-h and fdisk -1 to verify your work.

Ans. `fdisk -1 ; df -h`

7. Compare the output again of fdisk and df. Do both commands display the new partitions?

Ans. The newly created partitions are visible with fdisk.

But they are not displayed by df.

8. Create a backup with dd of the mbr that contains your 200MB primary partition.

Ans. dd if=/dev/sdc of=bootseccor.sdc.dd count=1 bs=512

9. Take a backup of the partition table containing your 400MB primary and 300MB logical drives. Make sure the logical drives are in the backup.

sfdisk -d /dev/sdb > parttable.sdb.sfdisk

10. List The filesysteins that are known by your system.

Ans. man fs

cat /proc/filesysterns

cat /etc/filesystems (not on all Linux distributions)

11. Create an ext2 filesystem on the 200MB partition.

Ans. mke2fs /dev/sdcl (replace sdcl with the correct partition)

12. Create an ext3 filesystem on the 400MB partition and one of the 300MB logic drives.

Ans. mke2fs -j /dev/sdbl (replace sdbl with the correct partition)

mke2fs -j /dev/sdb5 (replacce sdb5 with the correct partition)

13. Set the reserved space for root on the logical drive to 0 percent.

Ans. tune2fs -m 0 /dev/sdb5

14. Verify your work with fdisk and df.

Ans. mkfs (mke2fs) makes no difference in the output of these commands

The big change is in the next topic: mounting

EXERCISE

1. How many types of files there can be in UNIX?
2. How can security of files can be maintained?
3. What are different types of users in UNIX for files?
4. What is function of ls command? Give various options.
5. Differentiate between CP & MV Command.
6. Why is rm command used?
7. Differentiate between chmod & chown Command.
8. What do you understand by Editor?
9. How many types of Editors are present?
10. What is need of an editor?

❒❒❒

CHAPTER 6

CONFIGURING THE TCP/IP NETWORKING

Chapter Outline

- Network Layers
- Unicast, Multicast, Broadcast, Anycast
- LAN-WAN-MAN
- Internet-Intranet-Extranet
- TCP/IP
- To GUI or Not to GUI
- DEBIAN/UBUNTU NIC Configuration
- Setting the Host Name
- ifconfig
- Hostname
- ARP
- Route
- Ping
- Optional: Ethtool
- Mounting /proc File
- Assigning IP Addresses
- Creating Subnets
- Writing Hosts and Networks Files
- Interface Configuration for IP: Ifconfig
- Netstat Command
- Checking the ARP Tables; Name service and resolver configuration

6.1 NETWORK LAYERS: SEVEN OSI LAYERS

When talking about protocol layers, people usually mention the seven layers of the **osi protocol (Application, Presentation, Session, Transport, Network, Data Link and** Physical). We will discuss layers 2 and 3 in depth, and focus less on the other layers.

The reason is that these layers are important for understanding networks. You will hear administrators use words like "this is a layer 2 device" or "this is a layer 3 broadcast", and you should be able to understand what they are talking about.

four DoD layers

The DoD (or tcp/ip) model has only four layers, roughly mapping its network access **layer to OSI layers 1 and 2 (Physical and Datalink), its internet (IP) layer to the** OSI network layer, its host-to-host (tcp, udp) layer to OSI layer 4 (transport) and its application layer to OSI layers 5, 6 and 7.

Below an attempt to put OSI and DoD layers next to some protocols and devices.

physical layer

The physical layer, or layer 1, is all about voltage, electrical signals and mechanical connections. Some networks might still use coax cables, but most will have migrated to utp (cat 5 or better) with rj45 connectors.

Devices like repeaters and hubs are part of this layer. You cannot use software to 'see' a repeater or hub on the network. The only thing these devices are doing is amplifying electrical signals on cables. Passive hubs are multiport amplifiers that amplify an incoming electrical signal on all other connections. Active hubs do this by reading and retransmitting bits, without interpreting any meaning in those bits.

Network technologies like csma/cd and token ring are defined on this layer.

This is all we have to say about layer 1 in this book.

data link layer

The data link layer, or layer 2 is about frames. A frame has a crc (cyclic redundancy check). In the case of ethernet (802.3), each network card is identifiable by a unique 48-bit mac address (media access control address).

On this layer we find devices like bridges and switches. A bridge is more intelligent than a hub because a bridge can make decisions based on the mac address of computers. A switch also understands mac addresses.

In this book we will discuss commands like arp and ifconfig to explore this layer.

network layer

Layer 3 is about ip packets. This layer gives every host a unique 32-bit ip address. But ip is not the only protocol on this layer, there is also icmp, igmp, ipv6 and more. A complete list can be found in the /etc/protocols file.

On this layer we find devices like routers and layer 3 switches, devices that know (and have) an ip address.

In tcp/ip this layer is commonly referred to as the internet layer.

transport layer

We will discuss the tcp and udp protocols in the context of layer 4. The DoD model calls this the host-to-host layer.

layers 5, 6 and 7

The tcp/ip application layer includes layers 5, 6 and 7. Details on the difference between these layers are out of scope of this course.

[Protocols in Frame: eth:ip:udp:bootp]

Sniffing for ntp (Network Time Protocol) packets gives us this line, which makes us conclude to put ntp next to bootp in the protocol chart below.

[Protocols in Frame: eth:ip:udp:ntp]

Sniffing an arp broadcast makes us put arp next to ip. All these protocols are explained later in this chapter.

[Protocols in Frame: eth:arp]

Below is a protocol chart based on wireshark's knowledge. It contains some very common protocols that are discussed in this book. The chart does not contain all protocols.

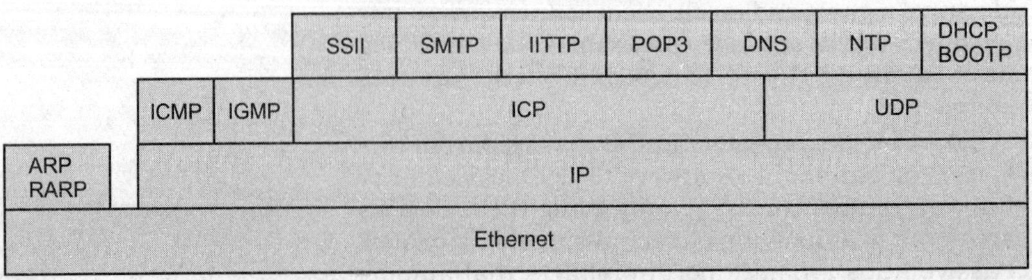

6.2 UNICAST, MULTICAST, BROADCAST, ANYCAST

unicast

A unicast communication originates from one computer and is destined for exactly one other computer (or host). It is common for computers to have many unicast communications.

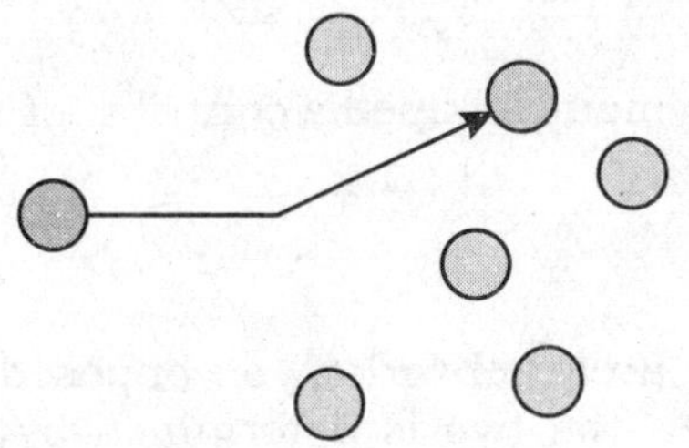

multicast

A multicast is destined for a group (of computers).

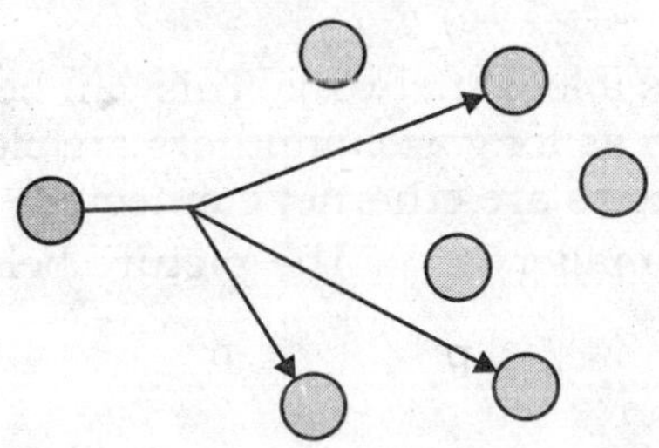

Some examples of multicast are Realplayer (.sdp files) and ripv2 (a routing protocol).

broadcast

A broadcast is meant for everyone.

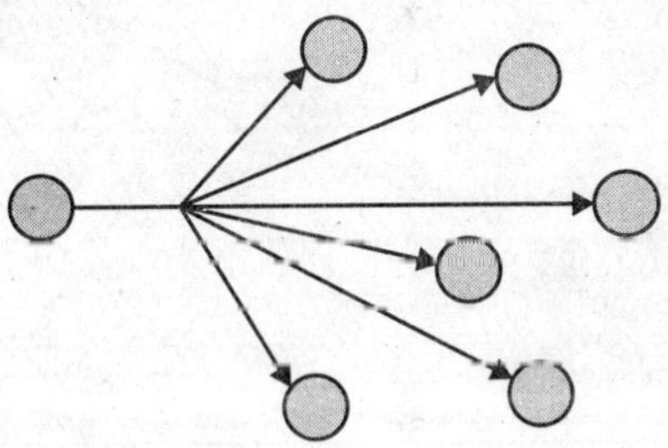

Typical example here is the BBC (British Broadcasting Corporation) broadcasting to everyone. In datacommunications a broadcast is most common confined to the lan.

Careful, a layer 2 broadcast is very different from a layer 3 broadcast. A layer two broadcast is received by all network cards on the same segment (it does not pass any router), whereas a layer 3 broadcast is received by all hosts in the same ip subnet.

anycast

The root name servers of the internet use anycast. An anycast signal goes the (geographically) nearest of a well defined group.

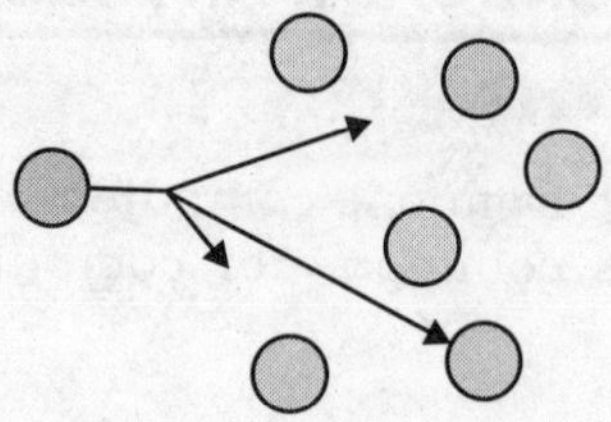

With thanks to the nice anonymous wikipedia contributor to put these pictures in the public domain.

6.3 LAN-WAN-MAN

The term lan is used for local area networks, as opposed to a wan for wide area networks. The difference between the two is determined by the distance between the computers, and not by the number of computers in a network. Some protocols like **atm are designed for use in a wan, others like ethernet are designed for use in a lan.**

lan

A lan (Local Area Network) is a local network. This can be one room, or one floor, or even one big building. We say lan as long as computers are close to each other. You can also define a lan when all computers are ethernet connected.

A lan can contain multiple smaller lan's. The picture below shows three lan's that together make up one lan.

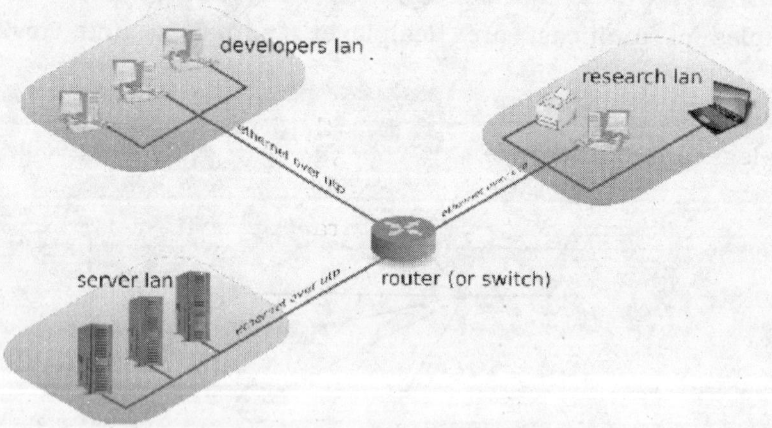

wan

A wan (Wide Area Network) is a network with a lot of distance between the computers (or hosts). These hosts are often connected by leased lines. A wan does not use ethernet, but protocols like fddi, frame relay, ATM or X.25 to connect computers (and networks).

The picture below shows a branch office that is connected through Frame Relay with headquarters.

The acronym wan is also, used for large surface area networks like the internet.

Cisco is known for their wan technology. They make routers that connect many lan networks using wan protocols.

man

A man (Metropolitan Area Network) is something in between a lan and a wan, often comprising several buildings on the same campus or in the same city. A man can use **fddi or ethernet or other protocols for connectivity.**

pan-wpan

Your home network is called a pan (Personal Area Network). A wireless pan is a **wpan.**

6.4 INTERNET-INTRANET-EXTRANET

The internet is a global network. It connects many networks using the tcp/ip protocol stack.

The origin of the internet is the arpanet. The arpanet was created in 1969, that year only four computers were connected in the network. In 1971 the first e-mail was sent over the arpanet. E-mail took 75 percent of all arpanet traffic in 1973. 1973 was also the year ftp was introduced, and saw the connection of the first European countries (Norway and UK). In 2009 the internet was available to 25 percent of the world population. In 2011 it is estimated that only a quarter of internet webpages are in English.

An intranet is a private tcp/ip network. An intranet uses the same protocols as the **internet, but is only accessible to people from within one organization.**

An extranet is similar to an intranet, but some trusted organizations (partners/ clients/suppliers/...) also get access.

6.5 TCP/IP

history of tcp/ip

In the sixties development of the tcp/ip protocol stack was started by the US Department of Defense. In the Eighties a lot of commercial enterprises developed their own protocol stack: IBM created sna, Novell had ipx/spx, Microsoft completed **netbeui and Apple worked with appletalk. All the efforts from the Eighties failed to** survive the Nineties. By the end of the Nineties, almost all computers in the world were able to speak tcp/ip.

In my humble opinion, the main reason for the survival of tcp/ip over all the other protocols is its openness. Everyone is free to develop and use the tcp/ip protocol suite.

rfc (request for comment)

The protocols that are used on the internet are defined in rfc's. An rfc or request for **comment describes the inner working of all internet protocols. The IETF (Internet** Engineering Task Force) is the sole publisher of these protocols since 1986.

The official website for the rfc's is http://www.rfc-editor.org. This website contains all rfc's in plain text, for example rfc2132 (which defines dhcp and bootp) is accessible at http://www.rfc-editor.org/rfc/rfc2132.txt.

many protocols

For reliable connections, you use tcp, whereas udp is connectionless but faster. The **icmp error messages are used by ping, multicast groups are managed by igmp.**

These protocols are visible in the protocol field of the ip header, and are listed in the **/etc/protocols file.**

```
paul@debian5:~$ grep tcp /etc/protocols
tcp   6     TCP   # transmission control protocol
```

many services

Network cards are uniquely identified by their mac address, hosts by their ip address and applications by their port number.

Common application level protocols like smtp, http, ssh, telnet and ftp have fixed **port numbers. There is a list of port numbers in /etc/services.**

```
paul@ubu1010:~$ grep ssh /etc/services
ssh     22/tcp          # SSH Remote Login Protocol
ssh     22/udp
```

6.6 TO GUI OR NOT TO GUI

Recent Linux distributions often include a graphical application to configure the network. Some people complain that these applications mess networking configurations up when used simultaneously with command line configurations. Notably Network Manager (often replaced by wicd) and yast are known to not care about configuration changes via the command line.

Since, the goal of this course is server administration, we will assume our Linux servers are always administered through the command line.

This chapter only focuses on using the command line for network interface configuration!

Unfortunately there is no single combination of Linux commands and/etc files that works on all Linux distributions. We discuss networking on two (large but distinct) Linux distribution families.

We start with Debian/Ubuntu, then continue with Fedora/RHEL.

6.7 DEBIAN/UBUNTU NIC CONFIGURATION

/etc/network/interfaces

The /etc/network/interfaces file is a core network interface card configuration file on Ubuntu and Debian.

dhcp client

The screenshot below shows that our current Ubuntu 11.04 is configured for dhcp on eth0 (the first network interface card or nic).

```
root@ubu1104srv:~# cat /etc/network/interfaces
# This file describes the network interfaces available on your system
# and how to activate them. For more information, see interfaces(5).

# The loopback network interface
auto lo
iface lo inet loopback

# The primary network interface
auto eth0
iface eth0 inet dhcp
```

Configuring network cards for dhcp is good practice for clients, but servers usually require a fixed ip address.

fixed ip

The screenshot below shows /etc/network/interfaces configured with a fixed ip **address.**

```
root@ubu1104srv:~# cat /etc/network/interfaces
# This file describes the network interfaces available on your system
# and how to activate them. For more information, see interfaces(5).

# The loopback network interface
auto lo
iface lo inet loopback

# The primary network interface
auto eth0
iface eth0 inet static
address 192.168.33.100
network 192.168.33.0
netmask 255.255.255.0
gateway 192.168.33.1
```

The screenshot above also shows that you can provide more configuration than just the ip address. See interfaces(5) for help on setting a gateway, netmask or any of the other options.

/sbin/ifdown

It is adviced (but not mandatory) to down an interface before changing its configuration. This can be done with the ifdown command.

The command will not give any output when downing an interface with a fixed ip address. However ifconfig will no longer show the interface.

```
root@ubu1104srv:~# ifdown eth0
root@ubu1104srv:~# ifconfig
lo     Link encap:Local Loopback
```

```
inet addr:127.0.0.1   Mask:255.0.0.0
inet6 addr: ::1/128 Scope:Host
UP LOOPBACK RUNNING  MTU:16436  Metric:1
RX packets:106 errors:0 dropped:0 overruns:0 frame:0
TX packets:106 errors:0 dropped:0 overruns:0 carrier:0
collisions:0 txqueuelen:0
RX bytes:11162 (11.1 KB)  TX bytes:11162 (11.1 KB)
```

An interface that is down cannot be used to connect to the network.

/sbin/ifup

Below a screenshot of ifup bringing the eth0 ethernet interface up using dhcp. (Note that this is a Ubuntu 10.10 screenshot, Ubuntu 11.04 omits ifup output by default.)

```
root@ubu1010srv:/etc/network# ifup eth0
Internet Systems Consortium DHCP Client V3.1.3
Copyright 2004-2009 Internet Systems Consortium.
All rights reserved.
For info, please visit https://www.isc.org/software/dhcp/

Listening on LPF/eth0/08:00:27:cd:7f:fc
Sending on LPF/eth0/08:00:27:cd:7f:fc
Sending on Socket/fallback
DHCPREQUEST of 192.168.1.34 on eth0 to 255.255.255.255 port 67
DHCPNAK from 192.168.33.100
DHCPDISCOVER on eth0 to 255.255.255.255 port 67 interval 3
DHCPOFFER of 192.168.33.77 from 192.168.33.100
DHCPREQUEST of 192.168.33.77 on eth0 to 255.255.255.255 port 67
DHCPACK of 192.168.33.77 from 192.168.33.100
bound to 192.168.33.77 - renewal in 95 seconds.
ssh stop/waiting
ssh start/running, process 1301
root@ubu1010srv:/etc/network#
```

The details of dhcp are covered in a separate chapter in the Linux Servers course.

6.8 SETTING THE HOST NAME

/etc/sysconfig/network

The /etc/sysconfig/network file is a global (across all network cards) configuration file. It allows us to define whether we want networking (NETWORKING=yes| no), what the hostname should be (HOSTNAME=) and which gateway to use (GATEWAY=).

```
[root@rhel6 ~]# cat /etc/sysconfig/network
NETWORKING=yes
HOSTNAME=rhel6
GATEWAY=192.168.1.1
```

There are a dozen more option settable in this file, details can be found in /usr/share/**doc/initscripts-*/sysconfig.txt.**

/etc/sysconfig/network-scripts/ifcfg-

Each network card can be configured individually using the /etc/sysconfig/network-**scripts/ifcfg-* files. When you have only one network card, then this will probably** be /etc/sysconfig/network-scripts/ifcfg-eth0.

dhcp client

Below a screenshot of /etc/sysconfig/network-scripts/ifcfg-eth0 configured for dhcp (BOOTPROTO="dhcp"). Note also the NM_CONTROLLED paramater to disable control of this nic by Network Manager. This parameter is not explained (not even mentioned) in /usr/share/doc/initscripts-*/sysconfig.txt, but many others are.

```
[root@rhel6 ~]# cat /etc/sysconfig/network-scripts/ifcfg-eth0
DEVICE="eth0"
HWADDR="08:00:27:DD:0D:5C"
NM_CONTROLLED="no"
BOOTPROTO="dhcp"
ONBOOT="yes"
```

The BOOTPROTO variable can be set to either dhcp or bootp, anything else will be considered static meaning there should be no protocol used at boot time to set the interface values.

fixed ip

Below a screenshot of a fixed ip configuration in /etc/sysconfig/network-scripts/**ifcfg-eth0.**

```
[root@rhel6 ~]# cat /etc/sysconfig/network-scripts/ifcfg-eth0
DEVICE="eth0"
HWADDR="08:00:27:DD:0D:5C"
NM_CONTROLLED="no"
BOOTPROTO="none"
IPADDR="192.168.1.99"
NETMASK="255.255.255.0"
GATEWAY="192.168.1.1"
ONBOOT="yes"
```

The HWADDR can be used to make sure that each network card gets the correct name when multiple network cards are present in the computer. It can not be used to assign a mac address to a network card. For this, you need to specify the MACADDR variable. Do not use HWADDR and MACADDR in the same ifcfg-ethx file.

The BROADCAST= and NETWORK= parameters from previous RHEL/Fedora versions are obsoleted.

/sbin/ifup and /sbin/ifdown

The ifup and ifdown commands will set an interface up or down, using the configuration discussed above. This is identical to their behaviour in Debian and Ubuntu.

```
[root@rhel6 ~]# ifdown eth0 && ifup eth0
[root@rhel6 ~]# ifconfig eth0
```

```
eth0  Link  encap:Ethernet   HWaddr  08:00:27:DD:0D:5C
      inet  addr:192.168.1.99  Bcast:192.168.1.255  Mask:255.255.255.0
      inet6  addr:  fe80::a00:27ff:fedd:d5c/64  Scope:Link
      UP  BROADCAST  RUNNING  MULTICAST  MTU:1500  Metric:1
      RX  packets:2452  errors:0  dropped:0  overruns:0  frame:0
      TX  packets:1881  errors:0  dropped:0  overruns:0  carrier:0
      collisions:0  txqueuelen:1000
      RX  bytes:257036  (251.0  KiB)  TX  bytes:184767  (180.4  KiB)
```

6.9 IFCONFIG

The use of /sbin/ifconfig without any arguments will present you with a list of all active network interface cards, including wireless and the loopback interface. In the screenshot below eth0 has no ip address.

```
root@ubu1010:~#  ifconfig
eth0  Link  encap:Ethernet   HWaddr  00:26:bb:5d:2e:52
      UP  BROADCAST  MULTICAST   MTU:1500   Metric:1
      RX  packets:0  errors:0  dropped:0  overruns:0  frame:0
      TX  packets:0  errors:0  dropped:0  overruns:0  carrier:0
      collisions:0  txqueuelen:1000
      RX  bytes:0  (0.0  B)   TX  bytes:0  (0.0  B)
      Interrupt:43  Base  address:0xe000

eth1  Link  encap:Ethernet   HWaddr  00:26:bb:12:7a:5e
      inet  addr:192.168.1.30   Bcast:192.168.1.255   Mask:255.255.255.0
      inet6  addr:  fe80::226:bbff:fe12:7a5e/64  Scope:Link
      UP  BROADCAST  RUNNING  MULTICAST   MTU:1500   Metric:1
      RX packets:11141791 errors:202 dropped:0 overruns:0 frame:11580126
      TX  packets:6473056  errors:3860  dropped:0  overruns:0  carrier:0
      collisions:0  txqueuelen:1000
      RX  bytes:3476531617  (3.4  GB)   TX  bytes:2114919475  (2.1  GB)
      Interrupt:23

lo    Link  encap:Local  Loopback
      inet  addr:127.0.0.1   Mask:255.0.0.0
      inet6  addr:  ::1/128  Scope:Host
      UP  LOOPBACK  RUNNING   MTU:16436   Metric:1
      RX  packets:2879  errors:0  dropped:0  overruns:0  frame:0
      TX  packets:2879  errors:0  dropped:0  overruns:0  carrier:0
      collisions:0  txqueuelen:0
      RX  bytes:486510  (486.5  KB)   TX  bytes:486510  (486.5  KB)
```

You can also use ifconfig to obtain information about just one network card.

```
[root@rhel6  ~]#  ifconfig  eth0
eth0  Link  encap:Ethernet   HWaddr  08:00:27:DD:0D:5C
      inet  addr:192.168.1.99   Bcast:192.168.1.255   Mask:255.255.255.0
      inet6  addr:  fe80::a00:27ff:fedd:d5c/64  Scope:Link
```

```
UP BROADCAST RUNNING MULTICAST  MTU:1500  Metric:1
RX packets:2969 errors:0 dropped:0 overruns:0 frame:0
TX packets:1918 errors:0 dropped:0 overruns:0 carrier:0
collisions:0 txqueuelen:1000
RX bytes:335942 (328.0 KiB)  TX bytes:190157 (185.7 KiB)
```

When /sbin is not in the $PATH of a normal user you will have to type the full path, as seen here on Debian.

```
paul@debian5:~$ /sbin/ifconfig eth3
eth3 Link encap:Ethernet  HWaddr 08:00:27:ab:67:30
     inet addr:192.168.1.29  Bcast:192.168.1.255  Mask:255.255.255.0
     inet6 addr: fe80::a00:27ff:feab:6730/64 Scope:Link
     UP BROADCAST RUNNING MULTICAST  MTU:1500  Metric:1
     RX packets:27155 errors:0 dropped:0 overruns:0 frame:0
     TX packets:30527 errors:0 dropped:0 overruns:0 carrier:0
     collisions:0 txqueuelen:1000
     RX bytes:13095386 (12.4 MiB)  TX bytes:25767221 (24.5 MiB)
```

up and down

You can also use ifconfig to bring an interface up or down. The difference with **ifup is that ifconfig eth0 up will re-activate the nic keeping its existing (current)** configuration, whereas ifup will read the correct file that contains a (possibly new) configuration and use this config file to bring the interface up.

```
[root@rhel6 ~]# ifconfig eth0 down
[root@rhel6 ~]# ifconfig eth0 up
[root@rhel6 ~]# ifconfig eth0
eth0 Link encap:Ethernet  HWaddr 08:00:27:DD:0D:5C
      inet addr:192.168.1.99  Bcast:192.168.1.255  Mask:255.255.255.0
      inet6 addr: fe80::a00:27ff:fedd:d5c/64 Scope:Link
      UP BROADCAST RUNNING MULTICAST  MTU:1500  Metric:1
      RX packets:2995 errors:0 dropped:0 overruns:0 frame:0
      TX packets:1927 errors:0 dropped:0 overruns:0 carrier:0
      collisions:0 txqueuelen:1000
      RX bytes:339030 (331.0 KiB)  TX bytes:191583 (187.0 KiB)
```

setting ip address

You can temporary set an ip address with ifconfig. This ip address is only valid until the next ifup/ifdown cycle or until the next reboot.

```
[root@rhel6 ~]# ifconfig eth0 | grep 192
      inet addr:192.168.1.99  Bcast:192.168.1.255  Mask:255.255.255.0
[root@rhel6 ~]# ifconfig eth0 192.168.33.42 netmask 255.255.0.0
[root@rhel6 ~]# ifconfig eth0 | grep 192
      inet addr:192.168.33.42  Bcast:192.168.255.255  Mask:255.255.0.0
[root@rhel6 ~]# ifdown eth0 && ifup eth0
[root@rhel6 ~]# ifconfig eth0 | grep 192
      inet addr:192.168.1.99  Bcast:192.168.1.255  Mask:255.255.255.0
```

setting mac address

You can also use ifconfig to set another mac address than the one hard coded in the network card. This screenshot shows you how.

```
[root@rhel6 ~]# ifconfig eth0 | grep HWaddr
eth0 Link encap:Ethernet  HWaddr 08:00:27:DD:0D:5C
[root@rhel6 ~]# ifconfig eth0 hw ether 00:42:42:42:42:42
[root@rhel6 ~]# ifconfig eth0 | grep HWaddr
eth0 Link encap:Ethernet  HWaddr 00:42:42:42:42:42
```

dhclient

Home and client Linux desktops often have /sbin/dhclient running. This is a daemon that enables a network interface to lease an ip configuration from a dhcp server.

When your adapter is configured for dhcp or bootp, then /sbin/ifup will start the **dhclient daemon.**

When a lease is renewed, dhclient will override your ifconfig set ip address!

6.10 HOSTNAME

Every host receives a hostname, often placed in a DNS name space forming the **fqdn or Fully Qualified Domain Name.**

This screenshot shows the hostname command and the configuration of the hostname on Red Hat/Fedora.

```
[root@rhel6 ~]# grep rhel /etc/sysconfig/network
HOSTNAME=rhel6
[root@rhel6 ~]# hostname
rhel6
```

Ubuntu/Debian uses the /etc/hostname file to configure the hostname.

```
paul@ubu1010:~$ cat /etc/hostname
ubu1010
paul@ubu1010:~$ hostname
ubu1010
```

On all Linux distributions you can change the hostname using the hostname **$newname command. This is not a permanent change.**

```
[root@rhel6 ~]# hostname server42
[root@rhel6 ~]# hostname
server42
```

On any Linux you can use sysctl to display and set the hostname.

```
[root@rhel6 ~]# sysctl kernel.hostname
kernel.hostname = server42
[root@rhel6 ~]# sysctl kernel.hostname=rhel6
kernel.hostname = rhel6
[root@rhel6 ~]# sysctl kernel.hostname
kernel.hostname = rhel6
```

```
[root@rhel6 ~]# hostname
rhel6
```

6.11 ARP

The ip to mac resolution is handled by the layer two broadcast protocol arp. The **arp table can be displayed with the arp tool. The screenshot below shows the list of** computers that this computer recently communicated with.

```
root@barry:~# arp -a
? (192.168.1.191) at 00:0C:29:3B:15:80 [ether] on eth1
agapi (192.168.1.73) at 00:03:BA:09:7F:D2 [ether] on eth1
anya (192.168.1.1) at 00:12:01:E2:87:FB [ether] on eth1
faith (192.168.1.41) at 00:0E:7F:41:0D:EB [ether] on eth1
kiss (192.168.1.49) at 00:D0:E0:91:79:95 [ether] on eth1
laika (192.168.1.40) at 00:90:F5:4E:AE:17 [ether] on eth1
pasha (192.168.1.71) at 00:03:BA:02:C3:82 [ether] on eth1
shaka (192.168.1.72) at 00:03:BA:09:7C:F9 [ether] on eth1
root@barry:~#
```

Anya is a Cisco Firewall, faith is a laser printer, kiss is a Kiss DP600, laika is a laptop and Agapi, Shaka and Pasha are SPARC servers. The question mark is a Red Hat Enterprise Linux server running on a virtual machine.

You can use arp -d to remove an entry from the arp table.

```
[root@rhel6 ~]# arp
Address          HWtype  HWaddress           Flags Mask            Iface
ubu1010          ether   00:26:bb:12:7a:5e   C                     eth0
anya             ether   00:02:cf:aa:68:f0   C                     eth0
[root@rhel6 ~]# arp -d anya
[root@rhel6 ~]# arp
Address          HWtype  HWaddress           Flags Mask            Iface
ubu1010          ether   00:26:bb:12:7a:5e   C                     eth0
anya             (incomplete)                                      eth0
[root@rhel6 ~]# ping anya
PING anya (192.168.1.1) 56(84) bytes of data.
64 bytes from anya (192.168.1.1): icmp_seq=1 ttl=254 time=10.2 ms
...
[root@rhel6 ~]# arp
Address          HWtype  HWaddress           Flags Mask            Iface
ubu1010          ether   00:26:bb:12:7a:5e   C                     eth0
anya             ether   00:02:cf:aa:68:f0   C                     eth0
```

6.12 ROUTE

You can see the computer's local routing table with the /sbin/route command (and also with netstat -r).

```
root@RHEL4b ~]# netstat -r
Kernel IP routing table
```

```
Destination      Gateway    Genmask  Flags    MSS Window  irtt Iface
192.168.1.0      *       255.255.255.0    U              0 0     0 eth0
[root@RHEL4b ~]# route
Kernel IP routing table
Destination      Gateway    Genmask        Flags Metric Ref Use Iface
192.168.1.0      * 255.255.255.0 U 0 0 0 eth0
[root@RHEL4b ~]#
```

It appears this computer does not have a gateway configured, so we use route add **default gw to add a default gateway on the fly.**

```
[root@RHEL4b ~]# route add default gw 192.168.1.1
[root@RHEL4b ~]# route
Kernel IP routing table
Destination      Gateway    Genmask      Flags Metric Ref Use Iface
192.168.1.0 * 255.255.255.0  U       0        0  0 eth0
default        192.168.1.1  0.0.0.0     UG      0   0   0 eth0
[root@RHEL4b ~]#
```

Unless you configure the gateway in one of the /etc/ file from the start of this chapter, your computer will forget this gateway after a reboot.

6.13 PING

If you can ping to another host, then tcp/ip is configured.

```
[root@RHEL4b ~]# ping 192.168.1.5
PING 192.168.1.5 (192.168.1.5) 56(84) bytes of data.
64 bytes from 192.168.1.5: icmp_seq=0 ttl=64 time=1004 ms
64 bytes from 192.168.1.5: icmp_seq=1 ttl=64 time=1.19 ms
64 bytes from 192.168.1.5: icmp_seq=2 ttl=64 time=0.494 ms
64 bytes from 192.168.1.5: icmp_seq=3 ttl=64 time=0.419 ms

- 192.168.1.5 ping statistics --
4 packets transmitted, 4 received, 0% packet loss, time 3009 ms
rtt min/avg/max/mdev = 0.419/251.574/1004.186/434.520 ms, pipe 2
[root@RHEL4b ~]#
```

6.14 OPTIONAL: ETHTOOL

To display or change network card settings, use ethtool. The results depend on the capabilities of your network card. The example shows a network that auto-negotiates it's bandwidth.

```
root@laika:~# ethtool eth0
Settings for eth0:
 Supported ports: [ TP ]
 Supported link modes: 10baseT/Half 10baseT/Full
                       100baseT/Half 100baseT/Full
                       1000baseT/Full
 Supports auto-negotiation: Yes
```

```
    Advertised link modes: 10baseT/Half 10baseT/Full
                           100baseT/Half 100baseT/Full
                           1000baseT/Full
    Advertised auto-negotiation: Yes
  Speed: 1000Mb/s
  Duplex: Full
  Port: Twisted Pair
  PHYAD: 0
  Transceiver: internal
  Auto-negotiation: on
  Supports Wake-on: pumbg
  Wake-on: g
  Current message level: 0x00000033 (51)
  Link detected: yes
```

This example shows how to use ethtool to switch the bandwidth from 1000 Mbit to 100 Mbit and back. Note that some time passes before the nic is back to 1000 Mbit.

```
root@laika:~# ethtool eth0 | grep Speed
Speed: 1000Mb/s
root@laika:~# ethtool -s eth0 speed 100
root@laika:~# ethtool eth0 | grep Speed
Speed: 100Mb/s
root@laika:~# ethtool -s eth0 speed 1000
root@laika:~# ethtool eth0 | grep Speed
Speed: 1000Mb/s
```

Questions

1. Verify whether dhclient is running.
2. Display your current ip address(es).
3. Display the configuration file where this ip address is defined.
4. Follow the nic configuration in the book to change your ip address from dhcp **client to fixed. Keep the same ip address to avoid conflicts!**
5. Did you also configure the correct gateway in the previous question ? If not, then do this now.
6. Verify that you have a gateway.
7. Verify that you can connect to the gateway, that it is alive.
8. Change the last two digits of your mac address.
9. Which ports are used by http, pop3, ssh, telnet, nntp and ftp ? Note that sctp was ommitted from the screenshot.
10. Explain why e-mail and websites are sent over tcp and not udp.
11. Display the hostname of your computer.
12. Which ip-addresses did your computer recently have contact with ?

solutions

1. Verify whether dhclient is running.

```
paul@debian5:~$ ps fax | grep dhclient
```

2. Display your current ip address(es).

```
paul@debian5:~$ /sbin/ifconfig | grep 'inet '
    inet addr:192.168.1.31   Bcast:192.168.1.255   Mask:255.255.255.0
    inet addr:127.0.0.1   Mask:255.0.0.0
```

3. Display the configuration file where this ip address is defined.

```
Ubuntu/Debian:  cat /etc/network/interfaces
Redhat/Fedora:  cat /etc/sysconfig/network-scripts/ifcfg-eth*
```

4. Follow the nic configuration in the book to change your ip address from dhcp **client to fixed. Keep the same ip address to avoid conflicts!**

```
Ubuntu/Debian:
ifdown eth0
vi /etc/network/interfaces
ifup eth0
Redhat/Fedora:
ifdown eth0
vi /etc/sysconfig/network-scripts/ifcfg-eth0
ifup eth0
```

5. Did you also configure the correct gateway in the previous question ? If not, then do this now.
6. Verify that you have a gateway.

```
paul@debian5:~$ /sbin/route
Kernel IP routing table
Destination     Gateway      Genmask        Flags Metric Ref   Use Iface
192.168.1.0     *            255.255.255.0  U     0      0     0   eth0
default         192.168.1.1  0.0.0.0        UG    0      0     0   eth0
```

7. Verify that you can connect to the gateway, that it is alive.

```
paul@debian5:~$ ping -c3 192.168.1.1
PING 192.168.1.1 (192.168.1.1) 56(84) bytes of data.
64 bytes from 192.168.1.1: icmp_seq=1 ttl=254 time=2.28 ms
64 bytes from 192.168.1.1: icmp_seq=2 ttl=254 time=2.94 ms
64 bytes from 192.168.1.1: icmp_seq=3 ttl=254 time=2.34 ms

- 192.168.1.1 ping statistics --
3 packets transmitted, 3 received, 0% packet loss, time 2008ms
rtt min/avg/max/mdev = 2.283/2.524/2.941/0.296 ms
```

8. Change the last two digits of your mac address.

```
[root@rhel6 ~]# ifconfig eth0 hw ether 08:00:27:ab:67:XX
```

9. Which ports are used by http, pop3, ssh, telnet, nntp and ftp ?

```
root@rhel6 ~# grep ^'http'  /etc/services
http     80/tcp   www www-http    # WorldWideWeb HTTP
http     80/udp   www www-http    # HyperText Transfer Protocol
```

```
root@rhel6 ~# grep ^'smtp ' /etc/services
smtp     25/tcp   mail
smtp     25/udp   mail
root@rhel6 ~# grep ^'ssh ' /etc/services
ssh      22/tcp                # The Secure Shell (SSH) Protocol
ssh      22/udp                # The Secure Shell (SSH) Protocol
root@rhel6 ~# grep ^'telnet ' /etc/services
telnet   23/tcp
telnet   23/udp
root@rhel6 ~# grep ^'nntp ' /etc/services
nntp     119/tcp  readnews untp  # USENET News Transfer Protocol
nntp     119/udp  readnews untp   # USENET News Transfer Protocol
root@rhel6 ~# grep ^'ftp ' /etc/services
ftp      21/tcp
ftp      21/udp   fsp fspd
```

Note that sctp was ommitted from the screenshot.

10. Explain why e-mail and websites are sent over tcp and not udp.

    Because tcp is reliable and udp is not.

11. Display the hostname of your computer.

```
paul@debian5:~$ hostname
debian5
```

12. Which ip-addresses did your computer recently have contact with ?

```
root@rhel6 ~# arp -a
? (192.168.1.1) at 00:02:cf:aa:68:f0 [ether] on eth2
? (192.168.1.30) at 00:26:bb:12:7a:5e [ether] on eth2
? (192.168.1.31) at 08:00:27:8e:8a:a8 [ether] on eth2
```

6.15 MOUNTING /PROC FILE

Mounting and unmounting: The file systems keep the files in a directory structure which is essentially a tree. So a new file system can be created by specifying the point of mount in the directory tree. A typical *mount* instruction has the following format.

mount a-block-special-file point-of-mount

Corresponding to a *mount* instruction, there is also an instruction to unmount. In Unix it is *mount* with the same format as *mount.*

In Unix every time we have a new disk added, it is mounted at a suitable point of mount in the directory tree. In that case the mount instruction is used exactly as explained. Of course, a disk is assumed to be formatted.

6.16 ASSIGNING IP ADDRESSES

The distribution-specific scripts and graphical tools are front-ends to ip (or ifconfig and route on older systems) to display and configure the kernel's networking configuration.

The ip command is used for assigning IP addresses to interfaces, for setting up routes to the Internet and to other networks, for displaying TCP/IP configurations etcetera.

The following commands show IP address and routing information:

```
bennyghome benny> ip addr show
1: lo: <LOOPBACK,UP> mtu 16436 qdisc noqueue
   link/loopback 00:00:00:00:00:00 brd 00:00:00:00:00:00
   inet 127.0.0.1/8 brd 127.255.255.255 scope host lo
   inet6 ::1/128 scope host
2: eth0: <BROADCAST, MULTICAST, UP> mtu 1500 qdisc pfifo_fast qlen 100
   link/ether 00:50:bf:7e:54:9a brd ff:ff:ff:ff:ff:ff
   inet 192.168.42.15/24 brd 192.168.42.255 scope global eth0
   inet6 fe80::250:bfff:fe7e:549a/10 scope link
benny@home benny> ip route show
192.168.42.0/24 dev eth0 scope link
127.0.0.0/8 dev lo scope link
default via 192.168.42.1 dev eth0
```

Things to note:

- two network interfaces, even on a system that has only one network interface card: "lo" is the local loop, used for internal network communication; "eth0" is a common name for a *real* interface. Do not ever change the local loop configuration, or your machine will start malfunctioning! Wireless interfaces are usually defined as "wlan0": modem interfaces as "ppp0". but there might be other names as well.
- IP addresses, marked with "inet": the local loop always has 127.0.0.1, the physical interface can have any other combination.
- The hardware address of your interface, which might be required as part of the authentication procedure to connect to a network, is marked with "ether". The local loop has 6 pairs of all zeros, the physical loop has 6 pairs of hexadecimal characters, of which the first 3 pairs are vendor-specific.

6.17 CREATING SUBNETS

Subnetting

In 1985, RFC 950 defined a standard procedure to support the subnetting, or division, of a single Class A. B, or C network number into smaller pieces. Subnetting was introduced to overcome some of the problems that parts of the Internet were beginning to experience with the classful two-level addressing hierarchy, such as:

- Internet routing tables were beginning to grow.
- Local administrators had to request another network number from the Internet before a new network could be installed at their site.

Both of these problems were attacked by adding another level of hierarchy to the IP addressing structure. Instead of the classful two-level hierarchy, subnetting supports a three-level hierarchy. Figure 7 Illustrates the basic Idea of subnetting. which Is to divide the standard classful host number field Into two parts-the subnet number and the host number on that subnet.

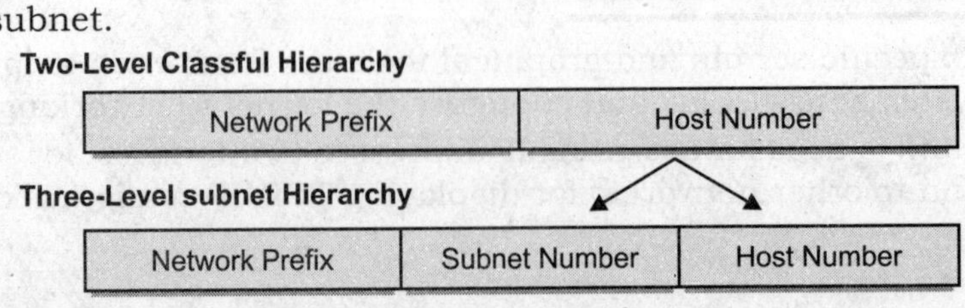

Subnetting attacked the expanding routing table problem by ensuring that the subnet structure of a network is never visible outside of the organization's private network. The route from the Internet to any subnet or a given IP address is the same, no matter which subnet the destination host is on. This is because all subnets of a given network number use the same network prefix but different subnet numbers. The routers within the private organization need to differentiate between the individual subnets, but as far as the Internet routers are concerned, all of the subnets in the organization are collected into a single routing table entry. This allows the local administrator to introduce arbitrary complexity Into the private network without affecting the size of the Internet's routing tables.

Subrentting overcame the registered number issue by assigning each organization one (or at most a few) network numbers from the IPv4 address space. The organization was then free to assign a distinct sub-network number for each of its internal networks. This allowed the organization to deploy additional subnets without obtaining a new network number from the Internet.

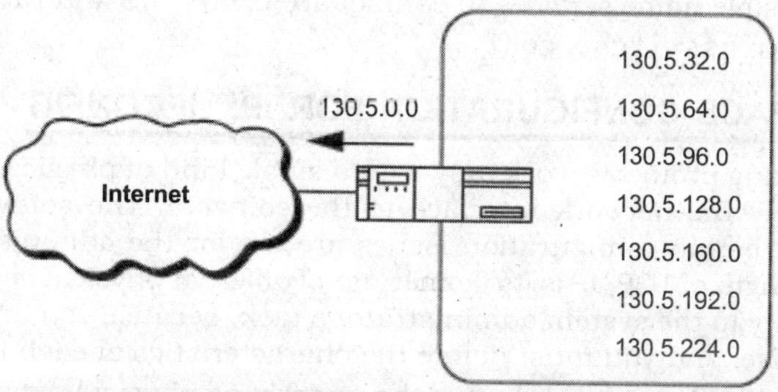

In Figure 8, a site with several logical networks uses subnet addressing with a single /16 (Class B) network address. The router accepts all traffic from the Internet addressed (o network 130.5.0.0, and forwards traffic to the interior subnetworks based on the third octet of the classful address. The deployment of subnetting within the private network provides several benefits:

- The size of the global Internet routing table does not grow because the site administrator does not need to obtain additional address space and the routing advertisements for all of the subnets arc combined into a single routing table entry.
- The local administrator has the flexibility to deploy additional subnets without obtaining a new network number from the Internet.
- Route flapping (that is, the rapid changing of routes) within the private network does not affect the Internet routing table since Internet routers do not know about the reachability of the individual subnets they Just know about the reachability of the parent network number.

6.18 WRITING HOSTS AND NETWORKS FILES

To display information on hosts or domains, use the host command;

```
[emmy@pc10 emmy]$ host www.Qunet.be
www.eunet.be. has address 193.74.208.177
[emmy/@pc10 emmy]$ host-t any Qunet.be
```

```
eunet.be.  SOA  dns.eunet.be.  hostmaster.Belgium.EU.net.
   2002021300  28800  7200  604800  86400
eunet.be.  mail  is  handled  by  50  pophost.eunet.be.
eunet.be  name  server  ns.EU.net.
eunet.be.  name  server  dns.eunet.be.
```

Similar, information can be displayed using the dig command, which give additional information about how records are stored in the name server.

The `/etc/nsswitch.conf` file defines the order in which to contact different name services. For Internet use, it is important that *dns* shows up in the "hosts" line:

```
[hob%tux~]  grep  hosts  /etc/nsswitch.conf
hosts:  files  dns
```

This instructs your computer to look up hostnames and IP addresses first in `/etc/hosts` file, and to contact the DNS server if a given host does not occur in the local `hosts` file. Other possible name services to contact are LDAP, NIS and NIS+.

More in man `nsswitch. conf.`

6.19 INTERFACE CONFIGURATION FOR IP: IFCONFIG

When networking protocols work only with a single kind of physical network, there is no need to identify the network interface to the software. The software knows what the interface *must* be; no configuration issues are left for the administrator. However, one important strength of TCP/IP is its flexible use of different physical networks. This flexibility adds complexity to the system administrator's task, because you must tell TCP/IP which interfaces to use, and you must define the characteristics of each interface.

Because TCP/IP is independent of the underlying physical network, IP addresses are implemented in the network software - not in the network hardware. Unlike Ethernet addresses, which are determined by the Ethernet hardware, the system administrator assigns an IP address to each network interface.

In this chapter, we use the **ifconfig** (interface configure) command to identify the network interface to TCP/IP and to assign the IP address, subnet mask, and broadcast address to the interface. We also configure a network interface to run Point-to-Point Protocol (PPP), which is the standard Network Access Layer protocol used to run TCP/IP over modem connections. Let's begin with a discussion of **ifconfig** .

6.19.1 The ifconfig Command

The **ifconfig** command sets, or checks, configuration values for network interfaces. Regardless of the vendor or version of UNIX, the **ifconfig** command will set the IP address, the subnet mask, and the broadcast address for each interface. Its most basic function is assigning the IP address.

Here is the **ifconfig** command that configures the Ethernet interface on *peanut* :

```
#
ifconfig  le0  172.16.12.2  netmask  255.255.255.0
\  broadcast  172.16.12.255
```

Many other arguments can be used with the **ifconfig** command; we discuss several of these later. But a few important arguments provide the basic information required by TCP/IP for every network interface. These are:

interface

The name of the network interface that you want to configure for TCP/IP. In the example above, this is the Ethernet interface le0.

address

The IP address assigned to this interface. Enter the address as either an IP address (in dotted decimal form) or as a hostname. If you use a hostname, place the hostname and its address in the */etc/hosts* file. Your system must be able to find the hostname in */etc/hosts* because **ifconfig** usually executes before DNS is running. The example uses the numeric IP address 172.16.12.2 as the **address** value.

netmask mask

The subnet mask for this interface. Ignore this argument only if you're using the default mask derived from the traditional address class structure. If you are subnetting, use your subnet mask. The subnet mask chosen for our imaginary network is 255.255.255.0, so that is the value assigned to *peanut* 's le0 interface. See chapters 2 and 4 for information on address masks and subnets.

broadcast address

The broadcast address for the network. Most, but not all, systems default to the standard broadcast address, which is an IP address with all host bits set to 1. In the **ifconfig** example we explicitly set the broadcast address to 172.16.12.255 to avoid any confusion. Every system on the subnet must agree on the broadcast address.

6.19.2 Determining the Interface Name

You can usually determine which interface is used on a system from the messages displayed on the console during a boot. On many systems these messages can be examined with the **dmesg** command. But even with this information, determining the name of the Ethernet interface is not always easy. The following example shows the output of the **dmesg** command on two different systems:

```
almond%
dmesg | grep le0
le0 at ledma0: SBus slot f 0xc00000 sparc ipl 6 le0 is /iommu@f,e0000000/
sbus@f,e0001000/ledma@f,400010/le@f,c00000   acorn>

dmesg | grep eth0
eth0: smc8432 (DEC 21041 Tulip) at 0xfc80, 00:00:c0:dd:d4:da, IRQ 10
eth0: enabling 10TP port.
```

The first **dmesg** command in the example shows the messages displayed when an le0 Ethernet interface is detected during the boot of a Solaris 2.5.1 system. Nothing about these messages makes it clear that le0 is an Ethernet interface. The second **dmesg** example, which comes from a PC running Linux, provides more clues. eth0 is a more intuitive Ethernet interface name; and the Linux system displays the Ethernet address (00:00:c0:dd:d4:da) and the make and model (SMC8432) of the network adapter card. If you know what these things mean, it makes guessing the Ethernet interface name simpler.

It is not always easy to determine all available interfaces on your system by looking at the output of **dmesg** , nor by looking at device statements in the kernel configuration file. These only show you the physical hardware interfaces. In the TCP/IP protocol architecture, the Network Access Layer encompasses all functions that fall below the Internet Layer. This can include all three lower layers of the OSI Reference Model: the Physical Layer, the Data Link Layer, and the Network Layer. IP needs to know the specific interface in the Network Access Layer where packets should be passed for delivery to a particular network. This interface is not limited to a physical hardware driver. It could be a software interface into the network layer of another protocol suite. So, what other methods can help you determine the network interfaces available on a system?

6.19.3 Checking the Interface with ifconfig

As noted above, the UNIX installation script configures the network interface. However, this configuration may not be exactly what you want. Check the configuration of an interface with **ifconfig**. To display the current values assigned to the interface, enter **ifconfig** with an interface name and no other arguments. For example, to check interface le0:

```
%
ifconfig le0
le0: flags=863<UP,BROADCAST,NOTRAILERS,RUNNING,MULTICAST> mtu 1500
inet 172.16.12.2 netmask ffff0000 broadcast 172.16.255.255
```

When used to check the status of an interface on a Solaris 2.5.1 system, the **ifconfig** command displays two lines of output. The first line shows the interface name, the flags that define the interface's characteristics, and the Maximum Transmission Unit (MTU) of this interface. In our example the interface name is le0, and the MTU is 1500 bytes. The flags are displayed as both a numeric value and a set of keywords. The interface's flags have the numeric value 863, which corresponds to:

UP

The interface is enabled for use.

BROADCAST

The interface supports broadcasts, which means it is connected to a network that supports broadcasts, such as an Ethernet.

NOTRAILERS

This interface does not support trailer encapsulation. This is an Ethernet-specific characteristic which we discuss in more detail later.

RUNNING

This interface is operational.

MULTICAST

This interface supports multicasting.

The second line of **ifconfig** output displays information that directly relates to TCP/IP. The keyword inet is followed by the Internet address assigned to this interface. Next comes the keyword netmask, followed by the address mask written in hexadecimal. Finally, the keyword broadcast and the broadcast address are displayed.

On a Linux system the **ifconfig** command displays up to six lines of information for each interface instead of the two lines displayed by the Solaris system. The additional information includes the Ethernet address, the PC IRQ and I/O Base Address, and packet statistics. The basic information is the same on both systems.

```
>
ifconfig  eth0
  eth0    Link  encap:10Mbps  Ethernet   HWaddr  00:00:C0:9A:D0:DB  inet
addr:172.16.55.106  Bcast:172.16.55.255  Mask:255.255.255.0 UP BROADCAST
RUNNING  MULTICAST   MTU:1500   Metric:1  RX  packets:844886  errors:0
dropped:0  overruns:0  TX  packets:7668  errors:0  dropped:0  overruns:0
Interrupt:11  Base  address:0x7c80
```

Refer to the Solaris **ifconfig le0** example at the beginning of this section. You'll see that the interface needs to be reconfigured. The configuration done by the user during the UNIX installation did not provide all of the values we planned. The address (172.16.12.2) is correct, but the address mask (ffff0000 or 255.255.0.0) and the broadcast address (172.16.0.0) are incorrect. Let's look at how these values are assigned, and how to correct them.

6.19.4 Assigning a Subnet Mask

In order to function properly, every interface on a specific physical network segment must have the same subnet mask. For le0 on *almond* and *peanut* , the netmask value is 255.255.255.0, because both systems are attached to the same subnet. However, although *almond* 's local network interface and its external network interface are parts of the same computer, they use different netmasks because they are on different networks.

To assign a subnet mask, write the subnet mask value after the keyword "netmask" on the **ifconfig** command line. The subnet mask is usually written in the "dotted decimal" form used for IP addresses. [3] For example, the following command assigns the correct subnet mask to the le0 interface on *peanut* :

[3] Hexadecimal notation can also be used for the subnet mask. To enter a netmask in hexadecimal form, write the value as a single hex number starting with a leading 0x. For example, the hexadecimal form of 255.255.255.0 is 0xffffff00. Choose the form that is easier for you to understand.

```
#
ifconfig le0 172.16.12.2 netmask 255.255.255.0\broadcast 172.16.12.255
```

Putting the netmask value directly on the **ifconfig** command line is the most common, the simplest, and the best way to manually assign the subnet mask to an interface. But it is also possible to tell **ifconfig** to take the netmask value from a file instead of from the command line. Conceptually, this is similar to using a hostname in place of an IP address. The administrator can place the subnet mask value in either the *hosts* file or the *networks* file and then reference it by name. For example, the *nuts-net* administrator might add the following entry to */etc/networks*:

```
nuts-mask          255.255.255.0
```

Once this entry has been added, you can use the name *nuts-mask* on the **ifconfig** command line, instead of the actual mask. For example:

```
#
ifconfig le0 172.16.5.2 netmask nuts-mask
```

The name *nuts-mask* resolves to 255.255.255.0, which is the correct netmask value for our sample systems.

On Solaris systems, you can also use */etc/inet/netmasks* to set the subnet mask. [4] The */etc/inet/netmasks* file is a table of one-line entries, each containing a network address separated from a subnet mask by whitespace. [5] If a Solaris system on *nuts-net* (172.16.0.0) has a */etc/inet/netmasks* file that contains the entry:

[4] */etc/netmasks* is symbolically linked to */etc/inet/netmasks* .

[5] Use the official network address, not a subnet address.

```
172.16.0.0          255.255.255.0
```

then the following **ifconfig** command can be used to set the subnet mask:

```
#
ifconfig le0 172.16.5.1 netmask +
```

The plus sign after the keyword netmask causes **ifconfig** to take the mask value from */etc/inet/netmasks*. **ifconfig** searches the file for a network address that matches the network address of the interface being configured. It then extracts the subnet mask associated with that address and applies it to the interface.

Some systems take advantage of the fact that the IP address, subnet mask, and broadcast address can be set indirectly to reduce the extent that startup files need to be customized. Reducing customization lessens the chance that a system might hang while booting because a startup file was improperly edited, and it makes it possible to pre-configure these files for all of the systems on the network. The *hosts*, *networks*, and *netmasks* files, which provide input to the **ifconfig** command, all produce NIS maps that can be centrally managed at sites using NIS.

A disadvantage of setting the **ifconfig** values indirectly is that it can make troubleshooting more cumbersome. If all values are set in the boot file, you only need to check the values there. When network configuration information is supplied indirectly, you may need to check the boot file, the *hosts* file, the *networks* file, and the *netmasks* file to find the problem. An error in any of these files could cause an incorrect configuration. To make debugging easier, many system administrators prefer to set the configuration values directly on the **ifconfig** command line.

Another disadvantage of setting the subnet mask value indirectly is that some of the files used for this are not primarily intended for this use. The *hosts* file is a particularly bad choice for storing subnet values. The *hosts* file is heavily used by other programs. Placing a subnet value in the *hosts* file might confuse one of these programs. Setting the subnet value directly on the command line or from a file, such as the *netmasks* file, that is dedicated to this purpose is probably the best approach.

6.19.5 Setting the Broadcast Address

RFC 919, *Broadcasting Internet Datagrams*, clearly defines the format of a broadcast address as an address with all host bits set to 1. Since, the broadcast address is so precisely defined, **ifconfig** should be able to compute it automatically, and you should always be able to use the default. Unfortunately, this is not the case. TCP/IP was included in BSD 4.2 before RFC 919 was an adopted standard. BSD 4.2 used a broadcast address with all host bits set to 0, and didn't allow the broadcast address to be modified during configuration. Because of this history, some releases of UNIX default to a "0-style" broadcast address for compatibility with older systems, while other releases default to the standard "1-style" broadcast address.

Avoid this confusion by defining a broadcast address for the entire network and ensuring that every device on the network explicitly sets it during configuration. Set the broadcast address in the **ifconfig** command using the keyword broadcast followed by the correct broadcast address. For example, the **ifconfig** command to set the broadcast address for *almond* 's le0 interface is:

```
#
ifconfig le0 172.16.12.1 netmask 255.255.255.0\broadcast 172.16.12.255
```

Note that the broadcast address is relative to the local subnet. *almond* views this interface as connected to network 172.16.12.0; therefore, its broadcast address is 172.16.12.255. Depending on the implementation, a UNIX system could interpret the address 172.16.255.255 as host address 255 on subnet 255 of network 172.16.0.0, or as the broadcast address for *nuts-net* as a whole. In neither case would it consider 172.16.255.255 the broadcast address for subnet 172.16.12.0.

6.19.6 The Other Command Options

We've used **ifconfig** to set the interface address, the subnet mask, and the broadcast address. These are certainly the most important functions of **ifconfig**, but it has other functions as well. It can enable or disable trailer encapsulation, the address resolution protocol, and the interface itself. **ifconfig** also can set the routing metric used by the Routing Information Protocol and the Maximum Transmission Unit (MTU) used by the interface. We'll look at each of these functions.

6.19.6.1 Enabling and disabling the interface

The **ifconfig** command has two arguments, **up** and **down**, for enabling and disabling the network interface. The **up** argument enables the network interface and marks it ready for use. The **down** argument disables the interface so that it cannot be used for network traffic.

Use the **down** argument when interactively reconfiguring an interface. Some configuration parameters - for example, the IP address - cannot be changed unless the interface is down. First, the interface is brought down. Then, the reconfiguration is done, and the interface is brought back up. For example, the following steps change the address for an interface:

```
#
ifconfig le0 down

 #

ifconfig le0 172.16.1.2 up
```

After these commands execute, the interface operates with the new configuration values. The **up** argument in the second **ifconfig** command is not actually required because it is the default. However, an explicit **up** is commonly used after the interface has been disabled, or when an **ifconfig** command is used in a script file to avoid problems if the default is changed in a future release.

6.19.6.2 Metric

On some systems, the **ifconfig** command creates an entry in the routing table for every interface that is assigned an IP address. Each interface is the route to a network. Even if a host isn't a gateway, its interface is still its "route" to the local network. **ifconfig** determines the route's destination network by applying the interface's subnet mask to the interface's IP address. For example, the le0 interface on *almond* has an address of 172.16.12.1 and a mask of 255.255.255.0. Applying this mask to the address provides the destination network, which is 172.16.12.0. The **netstat -in** display shows the destination address:

```
%
netstat  -in
Name Mtu  Net/Dest      Address       Ipkts    Ierrs Opkts Oerrs Collis
Queue le0  1500 172.16.12.0 172.16.12.1 1125826 16    569786  0    8914
0 lo0   1536 127.0.0.0       127.0.0.1    94280      0      94280 0 0    0
```

The Routing Information Protocol (RIP) is a routing protocol commonly used by UNIX. RIP does two things: it distributes routing information to other hosts, and it uses incoming routing information to build routing tables dynamically. The routes created by **ifconfig** are one source of the routing information distributed by RIP, and the **ifconfig metric** argument can be used to control how RIP uses this routing information.

RIP makes routing decisions based on the cost of a route. The route's cost is determined by a routing metric associated with the route. A routing metric is just a number. The lower the number, the lower the cost of the route. The higher the number, the higher the cost. When building a routing table, RIP favors low-cost routes over high-cost routes. Directly connected networks are given a very low cost. Therefore, the default metric is 0 for a route through an interface to a directly attached network. However, you can use the **metric** argument to supply a different routing metric for an interface.

To increase the cost of an interface to three, so that RIP prefers routes with values of 0, 1, or 2, use **metric 3** on the **ifconfig** command line:

```
#
ifconfig  std0  26.104.0.19  metric  3
```

Use the metric option only if there is another route to the same destination and you want to use it as the primary route. We did not use this command on *almond*, because it has only one interface connected to the outside world. But if it had a second connection, say, through a higher-speed link, then the command shown above could be used to direct traffic through the higher-performance interface.

A related **ifconfig** parameter is available on Solaris systems. RIP builds the routing table by choosing the most economical routes, and it distributes the routing table information to other hosts. The **metric** parameter controls which routes RIP selects as the lowest cost. The **private** argument, available on Solaris systems, controls the routes that RIP distributes. If **private** is specified on the **ifconfig** command line, the route created by that **ifconfig** command is not distributed by RIP. The default is **-private** , which permits the route to be distributed. The **private** parameter is not universally supported.

Additionally, not all systems make use of the **metric** argument. A Linux system doesn't create a routing table entry when it processes the **ifconfig** command. When configuring a Linux system, you add an explicit **route** command for each interface. (The **route** command is covered in the next chapter.) Linux systems ignore the **metric** argument.

Set the routing metric in a routing configuration file instead of on the **ifconfig** command line. This is the preferred method of providing routing information for newer routing software. We discuss the format of routing configuration files in the next chapter.

6.19.6.3 Maximum Transmission Unit

A network has a maximum transmission unit, which is the largest packet that can be transported over that physical network. On Ethernet, the maximum size is 1500 bytes, which is defined as part of the Ethernet standard. There is rarely any need to change the MTU on the **ifconfig** command line. By default, **ifconfig** chooses the optimum MTU, which is usually the largest legal MTU for a given type of network hardware. A large MTU is the default because it normally provides the best performance. However, a smaller MTU is helpful to achieve the following goals:

- To avoid fragmentation. If the traffic travels from a network with a large MTU - such as an FDDI network with an MTU of 4500 bytes - through a network with a smaller MTU like an Ethernet, the smaller MTU size may be best in order to avoid packet fragmentation. It is possible that specifying an MTU of 1500 on the interface connected to the FDDI may actually improve performance by avoiding fragmentation in the routers. This would only be done if fragmentation actually appeared to be the cause of a performance problem.
- To reduce buffer overruns or similar problems. On serial line connections it is possible to have equipment of such low performance that it cannot keep up with standard 1006-byte packets. In this case, it is possible to avoid buffer overruns or SILO overflows by using a smaller MTU. However, such solutions are temporary fixes. The real fix is to purchase the correct hardware for the application.

To change the MTU, use the **mtu** command-line argument:

```
# ifconfig fddi0 172.16.16.1 netmask 255.255.255.0 mtu 1500
```

This forces the FDDI interface on 172.16.16.1 to use an MTU of 1500 bytes.

6.19.6.4 Point-to-point

There are probably several more **ifconfig** command-line arguments available on your system. Linux has parameters to define the PC interrupt of the Ethernet hardware (**irq**) and the Ethernet hardware address (**hw**), and to enable multicasting (**multicast**) and promiscuous mode (**allmulti**). Solaris has arguments to set up or tear down the streams for an interface (**plumb/unplumb**), and to use Reverse ARP (RARP) to obtain the IP address for an interface (**auto-revarp**). But most of these parameters are not standardized between versions of UNIX.

One last feature that is available on most versions of UNIX is the ability to define point-to point connections with the **ifconfig** command. Point-to-point connections are network links that directly connect together only two computers. Of course the computers at either end of the link could be gateways to the world, but only two computers are directly connected to the link. Examples of a point-to-point connection are two computers linked together by a leased telephone line, or two computers in an office linked together by a null modem cable.

To define a point-to-point link on a Solaris system:

```
# ifconfig zs0 172.16.62.1 172.16.62.2
```

This **ifconfig** command has two addresses immediately following the interface name. The first is the address of the local host. The second address, called the destination address, is the address of the remote host at the other end of the point-to-point link. The second address shows up as the Net/Dest value in an **netstat -ni** display.

On a Linux system, this same configuration looks slightly different:

```
> ifconfig sl0 172.16.62.1 point-to-point 172.16.62.2
```

The syntax is different but the effect is the same. This enables the interface to run in point-to-point mode and identifies the hosts at both ends of the link.

Does this set up the Point-to-Point Protocol (PPP) used for TCP/IP serial line communication? No, it does not. These **ifconfig** parameters sometimes confuse people about how to set up PPP. There is much more to configuring PPP and SLIP, which we cover in the remainder of this chapter.

Before moving on to PPP and SLIP we should note that the configuration entered on an **ifconfig** command line will not survive a system boot. For a permanent configuration, put **ifconfig** in a startup file.

6.19.6.5 Putting ifconfig in the startup files

The **ifconfig** command is normally executed at boot time by a startup file. On BSD UNIX systems, the **ifconfig** commands are usually located in */etc/rc.boot* or */etc/rc.local.* System V UNIX presents a much more complex set of startup files, but the **ifconfig** statements are usually located in a file in the */etc/init.d* directory. [6] On Linux systems, the startup files containing the **ifconfig** commands are usually in the */etc/rc.d* or the */etc/rc.d/init.d* directory. Use **grep** to locate the specific file that contains the **ifconfig** command:

[6] A good description of the maze of System V initialization files is provided in *Essential System Administration, Second Edition* , by Æleen Frisch (O'Reilly & Associates).

```
> cd /etc/rc.d > grep ifconfig * rc.inet1:/sbin/ifconfig lo 127.0.0.1
rc.inet1:/sbin/ifconfig eth0 172.16.12.1 broadcast 172.16.12.255   netmask
255.255.255.0
```

Because network access is important for some of the processes run by the startup files, the **ifconfig** statements execute near the beginning of the startup procedure. The simplest way to configure a network interface to suit your requirements is to edit the startup files and insert the correct **ifconfig** statements.

On our Linux system, we edit */etc/rc.d/rc.inet1* and check the two lines necessary to configure the loopback interface and the Ethernet interface. We use the *rc.inet1* file because it runs early in network startup procedure and it is the file in which the installation script placed the **ifconfig** commands. The two lines placed in the file are:

```
ifconfig lo0 127.0.0.1 ifconfig eth0 172.16.12.1 broadcast 172.16.12.255
netmask 255.255.255.0
```

Check the startup files to ensure that the interfaces on the host are properly configured at every boot.

6.20 NETSTAT COMMAND

Use the **netstat** and the **ifconfig** commands. For example, to see all network interfaces that are already configured, enter:

```
%
netstat  -in
```

The **-i** option tells **netstat** to display the status of all configured network interfaces, and the **-n** tells **netstat** to display its output in numeric form. The **netstat -in** command displays the following fields:

Name

The Interface Name field shows the actual name assigned to the interface. This is the name you give to **ifconfig** to identify the interface. An asterisk (*) in this field indicates that the interface is not enabled; i.e., the interface is not "up."

Mtu

The Maximum Transmission Unit shows the longest frame (packet) that can be transmitted by this interface without fragmentation. The MTU is displayed in bytes. MTU is discussed in the section "

Net/Dest

The Network/Destination field shows the network or the destination host to which the interface provides access. In our Ethernet examples, this field contains a network address. The network address is derived from the IP address of the interface and the subnet mask. This field contains a host address if the interface is configured for a point-to-point (host-specific) link. The destination address is the address of the remote host at the other end of the point-to-point link. [1] A point-to-point link is a direct connection between two computers. You can create a point-to-point link with the **ifconfig** command. How this is done is covered later in this chapter.

Address

The IP Address field shows the Internet address assigned to this interface.

Ipkts

The Input Packets field shows how many packets this interface has received.

Ierrs

The Input Errors field shows how many damaged packets the interface has received.

Opkts

The Output Packets field shows how many packets were sent out by this interface.

Oerrs

The Output Errors field shows how many of the packets caused an error condition.

Collis

The Collisions field shows how many Ethernet collisions were detected by this interface. Ethernet collisions are a normal condition caused by Ethernet traffic contention. This field is not applicable to non-Ethernet interfaces.

Queue

The Packets Queued field shows how many packets are in the queue, awaiting transmission via this interface. Normally this is zero.

The output of a **netstat** command shows:

```
%
netstat  -in
 Name  Mtu     Net/Dest    Address          Ipkts Ierrs Opkts Oerrs Collis
Queue le0   1500   172.16.0.0 172.16.12.2 1547    1     1127   0   135
0 lo0     1536   127.0.0.0   127.0.0.1       133    0     133  0  0     0
```

This display shows that this workstation has only two network interfaces. In this case it is easy to identify each network interface. The lo0 interface is the loopback interface, which every TCP/IP system has

On most systems, the loopback interface is part of the default configuration, so you won't need to configure it. If you do need to configure lo0 on your system, use the following command:

```
#
ifconfig  lo0  127.0.0.1
```

The configuration of the Ethernet interface requires more attention. The surprising thing about the sample **netstat** display is that we haven't yet entered an **ifconfig** command for le0, and it already has an IP address! Many systems use an installation script to install UNIX. This script requests the host address, which it then uses to configure the interface. Later we'll look at whether the user successfully set up this interface with the installation script.

The **ifconfig** command can also be used to find out what network interfaces are available on a system. The **netstat** command shows only interfaces that are configured. On some systems the **ifconfig** command can be used to show all interfaces, even those that have not yet been configured. On Solaris 2.5.1 systems, **ifconfig -a** does this; on a Linux 2.0.0 system, entering **ifconfig** without any arguments will list all of the network interfaces.

While most hosts have only one real network interface, some hosts and all gateways have multiple interfaces. Sometimes all interfaces are the same type; i.e., a gateway between two Ethernets may have two Ethernet interfaces. **netstat** on a gateway like this might display lo0, le0, and le1. Deciphering a **netstat** display with multiple interfaces of the same type is still very simple. But deciphering a system with many different types of network interfaces is more difficult. You must rely on documentation that comes with optional software to choose the correct interface. When installing new network software, always read documentation carefully.

This long discussion about determining the network interface may seem to overshadow the important **ifconfig** functions of assigning the IP address, subnet mask, and broadcast address. So, let's return to these important topics.

6.21 CHECKING THE ARP TABLES; NAME SERVICE AND RESOLVER CONFIGURATION

Two options on the **ifconfig** command line, **arp** and **trailers** , are used only for Ethernet interfaces. The **trailers** option enables or disables negotiations for trailer encapsulation of IP packets. Trailer encapsulation is an optional technique that reduces the number of memory-to-memory copies the receiving system needs to perform.

To enable trailer encapsulation, put the keyword trailers on the **ifconfig** command line. When trailer encapsulation is enabled, the system requests (via the ARP protocol) that other systems also use trailer encapsulation when sending it data.

The option **-trailers** disables trailer encapsulation. Trailer encapsulation is disabled for two basic reasons. First, the I/O architecture of some systems does not derive any benefit from trailer encapsulation. If a system doesn't do memory-to-memory copies when receiving data from the network, it doesn't benefit from trailer encapsulation. Second, there are some systems that have difficulties with the negotiations for trailer encapsulation. For these reasons, many systems ignore the **trailers** argument and never use trailer encapsulation, and others allow trailer encapsulation but default to **-trailers** . Both of our sample systems, Solaris and Linux, ignore the **trailers** argument. However, some systems enable trailer encapsulation by default. Check your system documentation for the default on your system.

Enable ARP with the **ifconfig** keyword arp and disable it with the keyword -arp. It is possible (though very unlikely) that a host attached to your network cannot handle ARP. This would only happen on a network using specialized equipment or developmental hardware. In these very rare circumstances, it may be necessary to disable ARP in order to interoperate with the non-standard systems. By default, **ifconfig** enables ARP. Leave ARP enabled on all your systems.

SUMMARY

1. Most of the tasks covered in this chapter will generally have to be done only once. Afterward, you have to touch most configuration files only when adding a new system to your network or when you reconfigure your system entirely.
2. Some of the commands used to configure TCP/IP, however, have to be executed each time the system is booted. This is usually done by invoking them from the system */etc/rc** scripts.
3. Commonly, the network-specific part of this procedure is contained in a script. The name of this script varies in different Linux distributions.
4. In many older Linux distributions, it is known as *rc.net* or *rc.inet*. Sometimes you will also see two scripts named *rc.inet1* and *rc.inet2*; the former initializes the kernel part of networking and the latter starts basic networking services and applications.
5. In modern distributions, the *rc* files are structured in a more sophisticated arrangement; here you may find scripts in the */etc/init.d/* (or */etc/rc.d/init.d/*) directory that create the network devices and other *rc* files that run the network application programs.
6. You should then replace any sample commands in your configuration scripts with your commands, make sure the script is executed from the basic *rc* script at startup time, and reboot your machine.
7. The networking *rc* scripts that come along with your favorite Linux distribution should provide a solid example from which to work.
8. You are using one of the prepackaged Linux distributions, it will contain the major networking applications and utilities along with a coherent set of sample files. The only case in which you might have to obtain and install new utilities is when you install a new kernel release.
9. As they occasionally involve changes in the kernel networking layer, you will need to update the basic configuration tools.
10. This update at least involves recompiling, but sometimes you may also be required to obtain the latest set of binaries. These binaries are available at their official home site at ftp:/ ftp.inka.de/pub/comp/Linux/networking/NetTools/, packaged in an archive called *net-tools*

XXX.tar.gz, where *XXX* is the version number. The release matching Linux 2.0 is *net-tools 1.45.*

SHORT QUESTIONS WITH ANSWERS

1. Mount the small 20OMB partition on /home/project22.

Ans.
```
mkdir /home/project22
mount /dev/sdc1 /home/project22
```

2. Mount the big 40OMB primaiy partition on /mnt, the copy some files to it (everything in /etc). Then umount, and mount the file system as read only on /srv/ nfs/ salesnumbers. "Where are the files you copied ?

Ans.
```
mount /dev/sdb1 /mnt
cp -r /etc /mnt
ls -1 /mnt
umount /mnt
ls -1 /mnt
mkdir- -p /srv/nfs/salesnumbers
mount /dev/sdbl/srv/nfs/salesnumbers
You see the files in /srv/nfs/salenumbers now..
But physically they are on ext3 on partition /dev/sdb1
```

3. Verify your work with fdisk, df and mount. Also look in /etc/mtab and /proc/ mounts.

Ans.
```
fdisk -1
df -h
mount
All three the above commands should show your mounted partitions.
grep project22 /etc/mtab
grep project22 /proc/mounts
```

4. Make both mounts permanent, test that it works.

Ans.
```
add the following lines to /etc/fstab
/dev/sdc1 /home/project22 auto defaults 0 0
/dev/sdb1 /srv/nfs/salesnumbers auto defaults 0 0
```

5. What happens when you mount a file system on a directory that contains some files ?

Ans. `The files are hidden untill umount.`

6. What happens when you mount two file systems on the same mount point ?

Ans. `only the last mounted fs is visible.`

7. (optional) Describe the difference between these file searching commands: find, locate, updatedb, whereis, apropos and which.

Ans. `man is your friend`

8. (optional) Perform a file system check on the partition mounted at /srv/nfs/ salesnumbers.

9. Find the uuid of one of your cxt3 partitions with tune2fs and vol_id.

Ans.
```
root@rhel55:-# /lib/udev/vol_id -u /dev/hdal
60926898-2c78-49b4-a71d-c1d6310c87cc
root@ubul004:-# tune2fs -1 /dev/sda2 | grep UUID
Filesystem UUID: 3007b743-1dce-2d62-9a59-cf25f85191b7
```

10. Use this uuid in /etc/fstab and test that it works with a simple mount.

Ans.
```
tail -1 /etc/fstab
UUID=60926898-2c78-49b4-a71d-c1d6310c87cc /home/pro42 ext3 defaults 0 0
```

11. (optional) Test it also by removing a disk (so the device name is changed). You can edit settings in vmware/Virtualbox to remove a hard disk.

12. Display the root= directive in /boot/grub/menu.lst. (We see later in the course how to maintain this file.)

Ans.
```
paul@deb503:-$ grep ^[^#] /boot/grub/menu.1st | grep root=
kernel /boot/vmlinuz-2.6.26-2-686 root=/dev/hdal ro selinux=1 quiet
kernel /boot/vmlinuz-2.6.26-2-686 root=/dev/hdal ro selinux=1 single
```

13. Make a copy of the kernel, initrd and System.map tiles in /boot. Put the copies also in boot but replace 2.6.x with 3.0 (just imagine that Linux 3.0 is out.).

Ans.
```
cd /boot
cp vmlinuz-2.6.18-8.e15 vmlinuz-3.0
cp initrd-2.6.18-8.e15.img initra-3.0.img
cp System.map-2.6.18-8.e!5 System.roap-3.0
```

Do not forget the iuitrd file ends in .img.

14. Add a stanza in grub for the 3.0 files. Make sure the title is different.

Ans.
```
[root@RHEL5 ~]# grep 3.0 /boot/grub/menu.1st
title Reo Hat Enterprise Linux Server (3.0)
kernel /vmlimi2-3.0 ro root=/dev/VolGroup00/LogVol00 rhgb quiet
initrd /initrd-3.0.img
```

15. Set Ihe boot ineiiu liuieout to 30 seconds.

Ans.
```
[root@RHEL5 -]# grep time /boot/grub/menu.1st
timeout=30
```

EXERCISE

1. What is a filter?
2. Why are filters required?
3. How does a tee Command works?
4. Give syntex of head command
5. Give options of WC command
6. Why is cut command required.
7. Give uses of options of sort command.
8. What is a Shell Script?
9. What is similarity between DOS Batch file UNIX Shell script?
10. How many types of shell are present in UNIX environment
11. What is default shell for UNIX.
12. What is executable file name for Bourne Shell.

❐❐❐

CHAPTER 7

TCP/IP FIREWALL

7.0 INTRODUCTION

Security is increasingly important for companies and individuals alike. The Internet has provided them with a powerful tool to distribute information about themselves and obtain information from others, but it has also exposed them to dangers that they have previously been exempt from. Computer crime, information theft, and malicious damage are all potential dangers.

An unauthorized and unscrupulous person who gains access to a computer system may guess system passwords or exploit the bugs and idiosyncratic behavior of certain programs to obtain a working account on that machine. Once they are able to log in to the machine, they may have access to information that may be damaging, such as commercially sensitive information like marketing plans, new project details, or customer information databases. Damaging or modifying this type of data can cause severe setbacks to the company.

The safest way to avoid such widespread damage is to prevent unauthorized people from gaining network access to the machine. This is where firewalls come in.

Chapter Outline

- Introduction
- Methods of Attack
- What is a Firewall?
- What is IP Filtering?
- Setting Up Linux for Firewalling
- Testing a Firewall Configuration
- A Sample Firewall Configuration:
- IP Masquerade and Network Address Translation
- Side Effects and Fringe Benefits
- Configuring the Kernel for IP Masquerade
- Configuring IP Masquerade

7.1 METHODS OF ATTACK

As a network administrator, it is important that you understand the nature of potential attacks on computer security. We'll briefly describe the most important types of attacks so that you can better understand precisely what the Linux IP firewall will protect you against. You should do some additional reading to ensure that you are able to protect your network against other types of attacks. Here are some of the more important methods of attack and ways of protecting yourself against them:

Unauthorized access

This simply means that people who shouldn't use your computer services are able to connect and use them. For example, people outside your company might try to connect to your company accounting machine or to your NFS server.

There are various ways to avoid this attack by carefully specifying who can gain access through these services. You can prevent network access to all except the intended users.

Exploitation of known weaknesses in programs

Some programs and network services were not originally designed with strong security in mind and are inherently vulnerable to attack. The BSD remote services (rlogin, rexec, etc.) are an example.

The best way to protect yourself against this type of attack is to disable any vulnerable services or find alternatives. With Open Source, it is sometimes possible to repair the weaknesses in the software.

Denial of service

Denial of service attacks cause the service or program to cease functioning or prevent others from making use of the service or program. These may be performed at the network layer by sending carefully crafted and malicious datagrams that cause network connections to fail. They may also be performed at the application layer, where carefully crafted application commands are given to a program that cause it to become extremely busy or stop functioning.

Preventing suspicious network traffic from reaching your hosts and preventing suspicious program commands and requests are the best ways of minimizing the risk of a denial of service attack. It's useful to know the details of the attack method, so you should educate yourself about each new attack as it gets publicized.

Spoofing

This type of attack causes a host or application to mimic the actions of another. Typically the attacker pretends to be an innocent host by following IP addresses in network packets. For example, a well-documented exploit of the BSD rlogin service can use this method to mimic a TCP connection from another host by guessing TCP sequence numbers.

To protect against this type of attack, verify the authenticity of datagrams and commands. Prevent datagram routing with invalid source addresses. Introduce unpredictablility into connection control mechanisms, such as TCP sequence numbers and the allocation of dynamic port addresses.

Eavesdropping

This is the simplest type of attack. A host is configured to "listen" to and capture data not belonging to it. Carefully written eavesdropping programs can take usernames and passwords from user login network connections. Broadcast networks like Ethernet are especially vulnerable to this type of attack.

To protect against this type of threat, avoid use of broadcast network technologies and enforce the use of data encryption.

IP firewalling is very useful in preventing or reducing unauthorized access, network layer denial of service, and IP spoofing attacks. It not very useful in avoiding exploitation of weaknesses in network services or programs and eavesdropping.

7.2 WHAT IS A FIREWALL?

A firewall is a secure and trusted machine that sits between a private network and a public network . The firewall machine is configured with a set of rules that determine which network traffic will be allowed to pass and which will be blocked or refused. In some large organizations, you may even find a firewall located inside their corporate network to segregate sensitive areas of the organization from other employees. Many cases of computer crime occur from within an organization, not just from outside.

The term *firewall* comes from a device used to protect people from fire. The firewall is a shield of material resistant to fire that is placed between a potential fire and the people it is protecting.

Firewalls can be constructed in quite a variety of ways. The most sophisticated arrangement involves a number of separate machines and is known as a *perimeter network*. Two machines act as "filters" called chokes to allow only certain types of network traffic to pass, and between these chokes reside network servers such as a mail gateway or a World Wide Web proxy server. This configuration can be very safe and easily allows quite a great range of control over who can connect both from the inside to the outside, and from the outside to the inside. This sort of configuration might be used by large organizations.

Typically though, firewalls are single machines that serve all of these functions. These are a little less secure, because if there is some weakness in the firewall machine itself that allows people to gain access to it, the whole network security has been breached. Nevertheless, these types of firewalls are cheaper and easier to manage than the more sophisticated arrangement just described.

7.3 WHAT IS IP FILTERING?

IP filtering is simply a mechanism that decides which types of IP datagrams will be processed normally and which will be discarded. By *discarded* we mean that the datagram is deleted and completely ignored, as if it had never been received. You can apply many different sorts of criteria to determine which datagrams you wish to filter; some examples of these are:

- Protocol type: TCP, UDP, ICMP, etc.
- Socket number (for TCP/UPD)
- Datagram type: SYN/ACK, data, ICMP Echo Request, etc.
- Datagram source address: where it came from
- Datagram destination address: where it is going to

It is important to understand at this point that IP filtering is a network layer facility. This means it doesn't understand anything about the application using the network connections, only about the connections themselves. For example, you may deny users access to your internal network on the default telnet port, but if you rely on IP filtering alone, you can't stop them from using the telnet program with a port that you do allow to pass through your firewall. You can prevent this sort of problem by using proxy servers for each service that you allow across your firewall. The proxy servers understand the application they were designed to proxy and can therefore prevent abuses, such as using the telnet program to get past a firewall by using the World Wide Web port. If your firewall supports a World Wide Web proxy, their telnet connection will always be answered by the proxy and will allow only HTTP requests to pass. A large number of proxy-server programs exist. Some are free software and many others are commercial products. The Firewall-HOWTO discusses one popular set of these, but they are beyond the scope of this book.

The IP filtering ruleset is made up of many combinations of the criteria listed previously. For example, let's imagine that you wanted to allow World Wide Web users within the Virtual Brewery network to have no access to the Internet except to use other sites' web servers. You would configure your firewall to allow forwarding of:

- datagrams with a source address on Virtual Brewery network, a destination address of anywhere, and with a destination port of 80 (WWW)
- datagrams with a destination address of Virtual Brewery network and a source port of 80 (WWW) from a source address of anywhere.

Note that we've used two rules here. We have to allow our data to go out, but also the corresponding reply data to come back in. In practice, as we'll see shortly, Linux simplifies this and allows us to specify this in one command.

7.4 SETTING UP LINUX FOR FIREWALLING

To build a Linux IP firewall, it is necessary to have a kernel built with IP firewall support and the appropriate configuration utility. In all production kernels prior to the 2.2 series, you would use the ipfwadm utility. The 2.2.x kernels marked the release of the third generation of IP firewall for Linux called *IP Chains*. IP chains use a program similar to ipfwadm called ipchains. Linux kernels 2.3.15 and later support the fourth generation of Linux IP firewall called *netfilter*. The *netfilter* code is the result of a large redesign of the packet handling flow in Linux. The *netfilter* is a multifaceted creature, providing direct backward-compatible support for both ipfwadm and ipchains as well as a new alternative command called iptables. We'll talk about the differences between the three in the next few sections.

7.4.1 Kernel Configured with IP Firewall

The Linux kernel must be configured to support IP firewalling. There isn't much more to it than selecting the appropriate options when performing a make menuconfig of your kernel

```
Networking options ->
   [*] Network firewalls
   [*] TCP/IP networking
   [*] IP: firewalling
   [*] IP: firewall packet logging
```

Firewall packet logging is a special feature that writes a line of information about each datagram that matches a particular firewall rule out to a special device so you can see them.

In kernels 2.4.0 and later you should select this option instead:

```
Networking options ->
   [*] Network packet filtering (replaces ipchains)
   IP: Netfilter Configuration ->
      .
      <M> Userspace queueing via NETLINK (EXPERIMENTAL)
      <M> IP tables support (required for filtering/masq/NAT)
      <M> limit match support
      <M> MAC address match support
```

```
<M> netfilter MARK match support
<M> Multiple port match support
<M> TOS match support
<M> Connection state match support
<M> Unclean match support (EXPERIMENTAL)
<M> Owner match support (EXPERIMENTAL)
<M> Packet filtering
<M> REJECT target support
<M> MIRROR target support (EXPERIMENTAL)
.
<M> Packet mangling
<M> TOS target support
<M> MARK target support
<M> LOG target support
<M> ipchains (2.2-style) support
<M> ipfwadm (2.0-style) support
```

7.4.2 The ipfwadm Utility

The ipfwadm (IP Firewall Administration) utility is the tool used to build the firewall rules for all kernels prior to 2.2.0. Its command syntax can be very confusing because it can do such a complicated range of things, but we'll provide some common examples that will illustrate the most important variations of these.

The ipfwadm utility is included in most modern Linux distributions, but perhaps not by default. There may be a specific software package for it that you have to install. If your distribution does not include it, you can obtain the source package from *ftp.xos.nl* in the */pub/linux/ipfwadm/* directory, and compile it yourself.

7.4.3 The ipchains Utility

Just as for the ipfwadm utility, the ipchains utility can be somewhat baffling to use at first. It provides all of the flexibility of ipfwadm with a simplified command syntax, and additionally provides a "chaining" mechanism that allows you to manage multiple rulesets and link them together. We'll cover rule chaining in a separate section near the end of the chapter, because for most situations it is an advanced concept.

The ipchains command appears in most Linux distributions based on the 2.2 kernels Included in the source package is a wrapper script called ipfwadm-wrapper that mimics the ipfwadm command, but actually invokes the ipchains command. Migration of an existing firewall configuration is much more painless with this addition.

7.4.4 The iptables Utility

The syntax of the iptables utility is quite similar to that of the ipchains syntax. The changes are improvements and a result of the tool being redesigned to be extensible through shared libraries. Just as for ipchains, we'll present iptables equivalents of the examples so you can compare and contrast its syntax with the others.

7.5 TESTING A FIREWALL CONFIGURATION

Testing a Firewall Configuration

After you've designed an appropriate firewall configuration, it's important to validate that it does in fact do what you want it to do. One way to do this is to use a test host outside your network to attempt to pierce your firewall: this can be quite clumsy and slow, though, and is limited to testing only those addresses that you can actually use.

A faster and easier method is available with the Linux firewall implementation. It allows you to manually generate tests and run them through the firewall configuration just as if you were testing with actual datagrams. All varieties of the Linux kernel firewall software, ipfwadm, ipchains, and iptables, provide support for this style of testing. The implementation involves use of the relevant *check* command.

The general test procedure is as follows:

1. Design and configure your firewall using ipfwadm, ipchains, or iptables.
2. Design a series of tests that will determine whether your firewall is actually working as you intend. For these tests you may use any source or destination address, so choose some address combinations that should be accepted and some others that should be dropped. If you're allowing or disallowing only certain ranges of addresses, it is a good idea to test addresses on either side of the boundary of the range — one address just inside the boundary and one address just outside the boundary. This will help ensure that you have the correct boundaries configured, because it is sometimes easy to specify netmasks incorrectly in your configuration. If you're filtering by protocol and port number, your tests should also check all important combinations of these parameters. For example, if you intend to accept only TCP under certain circumstances, check that UDP datagrams are dropped.
3. Develop ipfwadm, ipchains, or iptables rules to implement each test. It is probably worthwhile to write all the rules into a script so you can test and re-test easily as you correct mistakes or change your design. Tests use almost the same syntax as rule specifications, but the arguments take on slightly differing meanings. For example, the source address argument in a rule specification specifies the source address that datagrams matching this rule should have. The source address argument in test syntax, in contrast, specifies the source address of the test datagram that will be generated. For ipfwadm, you must use the *-c* option to specify that this command is a test, while for ipchains and iptables, you must use the *-C* option. In all cases you must *always* specify the source address, destination address, protocol, and interface to be used for the test. Other arguments, such as port numbers or TOS bit settings, are optional.
4. Execute each test command and note the output. The output of each test will be a single word indicating the final target of the datagram after running it through the firewall configuration — that is, where the processing ended. For ipchains and iptables, user-specified chains will be tested in addition to the built-in ones.
5. Compare the output of each test against the desired result. If there are any discrepancies, you will need to analyse your ruleset to determine where you've made the error. If you've written your test commands into a script file, you can easily rerun the test after correcting any errors in your firewall configuration. It's a good practice to flush your rulesets completely and rebuild them from scratch, rather than to make changes dynamically. This helps ensure that the active configuration you are testing actually reflects the set of commands in your configuration script.

Let's take a quick look at what a manual test transcript would look like for our naïve example with ipchains. You will remember that our local network in the example was 172.16.1.0 with a netmask of 255.255.255.0, and we were to allow TCP connections out to web servers on the net. Nothing else was to pass our forward chain. Start with a transmission that we know should work, a connection from a local host to a web server outside:

```
# ipchains -C forward -p tcp -s 172.16.1.0 1025 -d 44.136.8.2 80
-i eth0
accepted
```

Note the arguments had to be supplied and the way they've been used to describe a datagram. The output of the command indicates that that the datagram was accepted for forwarding, which is what we hoped for.

Now try another test, this time with a source address that doesn't belong to our network. This one should be denied:

```
# ipchains -C forward -p tcp -s 172.16.2.0 1025 -d 44.136.8.2 80
-i eth0
denied
```

Try some more tests, this time with the same details as the first test, but with different protocols. These should be denied, too:

```
# ipchains -C forward -p udp -s 172.16.1.0 1025 -d 44.136.8.2 80
-i eth0
denied
# ipchains -C forward -p icmp -s 172.16.1.0 1025 -d 44.136.8.2 80
-i eth0
denied
```

Try another destination port, again expecting it to be denied:

```
# ipchains -C forward -p tcp -s 172.16.1.0 1025 -d 44.136.8.2 23
-i eth0
denied
```

You'll go a long way toward achieving peace of mind if you design a series of exhaustive tests. While this can sometimes be as difficult as designing the firewall configuration, it's also the best way of knowing that your design is providing the security you expect of it.

7.6 A SAMPLE FIREWALL CONFIGURATION:

A Sample Firewall Configuration

We've discussed the fundamentals of firewall configuration. Let's now look at what a firewall configuration might actually look like.

The configuration in this example has been designed to be easily extended and customized. We've provided three versions. The first version is implemented using the ipfwadm command (or the ipfwadm-wrapper script), the second uses ipchains, and the third uses iptables. The example doesn't attempt to exploit user-defined chains, but it will show you the similarities and differences between the old and new firewall configuration tool syntaxes:

```
#!/bin/bash
##########################################################################
# IPFWADM VERSION
```

```
# This sample configuration is for a single host firewall configuration
# with no services supported by the firewall machine itself.
##########################################################################
# USER CONFIGURABLE SECTION
# The name and location of the ipfwadm utility. Use ipfwadm-wrapper
for
# 2.2.* kernels.
IPFWADM=ipfwadm

# The path to the ipfwadm executable.
PATH="/sbin"

# Our internal network address space and its supporting network
device.
OURNET="172.29.16.0/24"
OURBCAST="172.29.16.255"
OURDEV="eth0"

# The outside address and the network device that supports it.
ANYADDR="0/0"
ANYDEV="eth1"

# The TCP services we wish to allow to pass - "" empty means all
ports
# note: space separated
TCPIN="smtp www"
TCPOUT="smtp www ftp ftp-data irc"

# The UDP services we wish to allow to pass - "" empty means all
ports
# note: space separated
UDPIN="domain"
UDPOUT="domain"

# The ICMP services we wish to allow to pass - "" empty means all
types
# ref: /usr/include/netinet/ip_icmp.h for type numbers
# note: space separated
ICMPIN="0 3 11"
ICMPOUT="8 3 11"

# Logging; uncomment the following line to enable logging of datagrams
# that are blocked by the firewall.
# LOGGING=1

# END USER CONFIGURABLE SECTION
```

```
########################################################################
# Flush the Incoming table rules
$IPFWADM -I -f

# We want to deny incoming access by default.
$IPFWADM -I -p deny

# SPOOFING
# We should not accept any datagrams with a source address matching
ours
# from the outside, so we deny them.
$IPFWADM -I -a deny -S $OURNET -W $ANYDEV

# SMURF
# Disallow ICMP to our broadcast address to prevent "Smurf" style
attack.
$IPFWADM -I -a deny -P icmp -W $ANYDEV -D $OURBCAST

# TCP
# We will accept all TCP datagrams belonging to an existing connection
# (i.e. having the ACK bit set) for the TCP ports we're allowing
through.
# This should catch more than 95 % of all valid TCP packets.
$IPFWADM -I -a accept -P tcp -D $OURNET $TCPIN -k -b

# TCP - INCOMING CONNECTIONS
# We will accept connection requests from the outside only on the
# allowed TCP ports.
$IPFWADM -I -a accept -P tcp -W $ANYDEV -D $OURNET $TCPIN -y

# TCP - OUTGOING CONNECTIONS
# We accept all outgoing tcp connection requests on allowed TCP ports.
$IPFWADM -I -a accept -P tcp -W $OURDEV -D $ANYADDR $TCPOUT -y

# UDP - INCOMING
# We will allow UDP datagrams in on the allowed ports.
$IPFWADM -I -a accept -P udp -W $ANYDEV -D $OURNET $UDPIN

# UDP - OUTGOING
# We will allow UDP datagrams out on the allowed ports.
$IPFWADM -I -a accept -P udp -W $OURDEV -D $ANYADDR $UDPOUT

# ICMP - INCOMING
# We will allow ICMP datagrams in of the allowed types.
$IPFWADM -I -a accept -P icmp -W $ANYDEV -D $OURNET $UDPIN
```

```
# ICMP - OUTGOING
# We will allow ICMP datagrams out of the allowed types.
$IPFWADM -I -a accept -P icmp -W $OURDEV -D $ANYADDR $UDPOUT

# DEFAULT and LOGGING
# All remaining datagrams fall through to the default
# rule and are dropped. They will be logged if you've
# configured the LOGGING variable above.
#
if [ "$LOGGING" ]
then
   # Log barred TCP
   $IPFWADM -I -a reject -P tcp -o

   # Log barred UDP
   $IPFWADM -I -a reject -P udp -o

   # Log barred ICMP
   $IPFWADM -I -a reject -P icmp -o
fi
#
# end.
```

Now we'll reimplement it using the ipchains command:

```
#!/bin/bash
##########################################################################
# IPCHAINS VERSION
# This sample configuration is for a single host firewall configuration
# with no services supported by the firewall machine itself.
##########################################################################

# USER CONFIGURABLE SECTION

# The name and location of the ipchains utility.
IPCHAINS=ipchains

# The path to the ipchains executable.
PATH="/sbin"

# Our internal network address space and its supporting network
device.
OURNET="172.29.16.0/24"
OURBCAST="172.29.16.255"
OURDEV="eth0"

# The outside address and the network device that supports it.
```

```
ANYADDR="0/0"
ANYDEV="eth1"

# The TCP services we wish to allow to pass - "" empty means all
ports
# note: space separated
TCPIN="smtp www"
TCPOUT="smtp www ftp ftp-data irc"

# The UDP services we wish to allow to pass - "" empty means all
ports
# note: space separated
UDPIN="domain"
UDPOUT="domain"

# The ICMP services we wish to allow to pass - "" empty means all
types
# ref: /usr/include/netinet/ip_icmp.h for type numbers
# note: space separated
ICMPIN="0 3 11"
ICMPOUT="8 3 11"

# Logging; uncomment the following line to enable logging of datagrams
# that are blocked by the firewall.
# LOGGING=1

# END USER CONFIGURABLE SECTION
###########################################################################
# Flush the Input table rules
$IPCHAINS -F input

# We want to deny incoming access by default.
$IPCHAINS -P input deny

# SPOOFING
# We should not accept any datagrams with a source address matching
ours
# from the outside, so we deny them.
$IPCHAINS -A input -s $OURNET -i $ANYDEV -j deny

# SMURF
# Disallow ICMP to our broadcast address to prevent "Smurf" style
attack.
$IPCHAINS -A input -p icmp -w $ANYDEV -d $OURBCAST -j deny

# We should accept fragments, in ipchains we must do this explicitly.
```

```
$IPCHAINS -A input -f -j accept

# TCP
# We will accept all TCP datagrams belonging to an existing connection
# (i.e. having the ACK bit set) for the TCP ports we're allowing
through.
# This should catch more than 95 % of all valid TCP packets.
$IPCHAINS -A input -p tcp -d $OURNET $TCPIN ! -y -b -j accept

# TCP - INCOMING CONNECTIONS
# We will accept connection requests from the outside only on the
# allowed TCP ports.
$IPCHAINS -A input -p tcp -i $ANYDEV -d $OURNET $TCPIN -y -j accept

# TCP - OUTGOING CONNECTIONS
# We accept all outgoing TCP connection requests on allowed TCP ports.
$IPCHAINS -A input -p tcp -i $OURDEV -d $ANYADDR $TCPOUT -y -j accept

# UDP - INCOMING
# We will allow UDP datagrams in on the allowed ports.
$IPCHAINS -A input -p udp -i $ANYDEV -d $OURNET $UDPIN -j accept

# UDP - OUTGOING
# We will allow UDP datagrams out on the allowed ports.
$IPCHAINS -A input -p udp -i $OURDEV -d $ANYADDR $UDPOUT -j accept

# ICMP - INCOMING
# We will allow ICMP datagrams in of the allowed types.
$IPCHAINS -A input -p icmp -w $ANYDEV -d $OURNET $UDPIN -j accept

# ICMP - OUTGOING
# We will allow ICMP datagrams out of the allowed types.
$IPCHAINS -A input -p icmp -i $OURDEV -d $ANYADDR $UDPOUT -j accept

# DEFAULT and LOGGING
# All remaining datagrams fall through to the default
# rule and are dropped. They will be logged if you've
# configured the LOGGING variable above.
#
if [ "$LOGGING" ]
then
   # Log barred TCP
   $IPCHAINS -A input -p tcp -l -j reject

   # Log barred UDP
   $IPCHAINS -A input -p udp -l -j reject
```

```
    # Log barred ICMP
    $IPCHAINS -A input -p icmp -l -j reject
fi
#
# end.
```

In our iptables example, we've switched to using the FORWARD ruleset because of the difference in meaning of the INPUT ruleset in the *netfilter* implementation. This has implications for us; it means that none of the rules protect the firewall host itself. To accurately mimic our ipchains example, we would replicate each of our rules in the INPUT chain. For clarity, we've dropped all incoming datagrams received from our outside interface instead.

```
#!/bin/bash
##########################################################################
# IPTABLES VERSION
# This sample configuration is for a single host firewall configuration
# with no services supported by the firewall machine itself.
##########################################################################

# USER CONFIGURABLE SECTION

# The name and location of the ipchains utility.
IPTABLES=iptables

# The path to the ipchains executable.
PATH="/sbin"

# Our internal network address space and its supporting network device.
OURNET="172.29.16.0/24"
OURBCAST="172.29.16.255"
OURDEV="eth0"

# The outside address and the network device that supports it.
ANYADDR="0/0"
ANYDEV="eth1"

# The TCP services we wish to allow to pass - "" empty means all ports
# note: comma separated
TCPIN="smtp,www"
TCPOUT="smtp,www,ftp,ftp-data,irc"

# The UDP services we wish to allow to pass - "" empty means all ports
# note: comma separated
UDPIN="domain"
UDPOUT="domain"
```

```
# The ICMP services we wish to allow to pass - "" empty means all
types
# ref: /usr/include/netinet/ip_icmp.h for type numbers
# note: comma separated
ICMPIN="0,3,11"
ICMPOUT="8,3,11"

# Logging; uncomment the following line to enable logging of datagrams
# that are blocked by the firewall.
# LOGGING=1

# END USER CONFIGURABLE SECTION
##########################################################################
# Flush the Input table rules
$IPTABLES -F FORWARD

# We want to deny incoming access by default.
$IPTABLES -P FORWARD deny

# Drop all datagrams destined for this host received from outside.
$IPTABLES -A INPUT -i $ANYDEV -j DROP

# SPOOFING
# We should not accept any datagrams with a source address matching
ours
# from the outside, so we deny them.
$IPTABLES -A FORWARD -s $OURNET -i $ANYDEV -j DROP

# SMURF
# Disallow ICMP to our broadcast address to prevent "Smurf" style
attack.
$IPTABLES -A FORWARD -m multiport -p icmp -i $ANYDEV -d $OURNET -j
DENY

# We should accept fragments, in iptables we must do this explicitly.
$IPTABLES -A FORWARD -f -j ACCEPT

# TCP
# We will accept all TCP datagrams belonging to an existing connection
# (i.e. having the ACK bit set) for the TCP ports we're allowing
through.
# This should catch more than 95 % of all valid TCP packets.
$IPTABLES -A FORWARD -m multiport -p tcp -d $OURNET –dports $TCPIN /
   ! –tcp-flags SYN,ACK ACK -j ACCEPT
$IPTABLES -A FORWARD -m multiport -p tcp -s $OURNET –sports $TCPIN /
   ! –tcp-flags SYN,ACK ACK -j ACCEPT
```

```
# TCP - INCOMING CONNECTIONS
# We will accept connection requests from the outside only on the
# allowed TCP ports.
$IPTABLES -A FORWARD -m multiport -p tcp -i $ANYDEV -d $OURNET $TCPIN
/ -syn -j ACCEPT

# TCP - OUTGOING CONNECTIONS
# We will accept all outgoing tcp connection requests on the allowed
/
TCP ports.
$IPTABLES -A FORWARD -m multiport -p tcp -i $OURDEV -d $ANYADDR /
   -dports $TCPOUT -syn -j ACCEPT
# UDP - INCOMING
# We will allow UDP datagrams in on the allowed ports and back.
$IPTABLES -A FORWARD -m multiport -p udp -i $ANYDEV -d $OURNET /
   -dports $UDPIN -j ACCEPT
$IPTABLES -A FORWARD -m multiport -p udp -i $ANYDEV -s $OURNET /
   -sports $UDPIN -j ACCEPT
# UDP - OUTGOING
# We will allow UDP datagrams out to the allowed ports and back.
$IPTABLES -A FORWARD -m multiport -p udp -i $OURDEV -d $ANYADDR /
   -dports $UDPOUT -j ACCEPT
$IPTABLES -A FORWARD -m multiport -p udp -i $OURDEV -s $ANYADDR /
   -sports $UDPOUT -j ACCEPT
# ICMP - INCOMING
# We will allow ICMP datagrams in of the allowed types.
$IPTABLES -A FORWARD -m multiport -p icmp -i $ANYDEV -d $OURNET /
   -dports $ICMPIN -j ACCEPT
# ICMP - OUTGOING
# We will allow ICMP datagrams out of the allowed types.
$IPTABLES -A FORWARD -m multiport -p icmp -i $OURDEV -d $ANYADDR /
   -dports $ICMPOUT -j ACCEPT
# DEFAULT and LOGGING
# All remaining datagrams fall through to the default
# rule and are dropped. They will be logged if you've
# configured the LOGGING variable above.
#
if [ "$LOGGING" ]
then
   # Log barred TCP
   $IPTABLES -A FORWARD -m tcp -p tcp -j LOG
   # Log barred UDP
   $IPTABLES -A FORWARD -m udp -p udp -j LOG
   # Log barred ICMP
   $IPTABLES -A FORWARD -m udp -p icmp -j LOG
fi
```

```
#
# end.
```

In many simple situations, to use the sample all you have to do is edit the top section of the file labeled “USER CONFIGURABLE section” to specify which protocols and datagrams type you wish to allow in and out. For more complex configurations, you will need to edit the section at the bottom, as well. Remember, this is a simple example, so scrutinize it very carefully to ensure it does what you want while implementing it.

7.6.1 IPAccounting,

In today’s world of commercial Internet service, it is becoming increasingly important to know how much data you are transmitting and receiving on your network connections. If you are an Internet Service Provider and you charge your customers by volume, this will be essential to your business. If you are a customer of an Internet Service Provider that charges by data volume, you will find it useful to collect your own data to ensure the accuracy of your Internet charges.

There are other uses for network accounting that have nothing to do with dollars and bills. If you manage a server that offers a number of different types of network services, it might be useful to you to know exactly how much data is being generated by each onc. This sort of information could assist you in making decisions, such as what hardware to buy or how many servers to run.

The Linux kernel provides a facility that allows you to collect all sorts of useful information about the network traffic it sees. This facility is called *IP accounting.*

7.6.2 Configuring the Kernel for IP Accounting

The Linux IP accounting feature is very closely related to the Linux firewall software. The places you want to collect accounting data are the same places that you would be interested in performing firewall filtering: into and out of a network host, and in the software that does the routing of datagrams. If you haven’t read the section on firewalls, now is probably a good time to do so.

To activate the Linux IP accounting feature, you should first see if your Linux kernel is configured for it. Check to see if the */proc/net/ip_acct* file exists. If it does, your kernel already supports IP accounting. If it doesn’t, you must build a new kernel, ensuring that you answer “Y” to the options in 2.0 and 2.2 series kernels:

```
Networking options ->
    [*] Network firewalls
    [*] TCP/IP networking
    ...
    [*] IP: accounting
```

or in 2.4 series kernels:

```
Networking options ->
    [*] Network packet filtering (replaces ipchains)
```

7.6.3 Configuring IP Accounting

Because IP accounting is closely related to IP firewall, the same tool was designated to configure it, so ipfwadm, ipchains or iptables are used to configure IP accounting. The command syntax is very similar to that of the firewall rules, so we won’t focus on it, but we will discuss what you can discover about the nature of your network traffic using this feature.

The general syntax for IP accounting with ipfwadm is:

```
# ipfwadm -A [direction] [command] [parameters]
```

The direction argument is new. This is simply coded as in, out, or both. These directions are from the perspective of the linux machine itself, so in means data coming into the machine from a network connection and out means data that is being transmitted by this host on a network connection. The both direction is the sum of both the incoming and outgoing directions.

The general command syntax for ipchains and iptables is:

```
# ipchains -A chain rule-specification
# iptables -A chain rule-specification
```

The ipchains and iptables commands allow you to specify direction in a manner more consistent with the firewall rules. IP Firewall Chains doesn't allow you to configure a rule that aggregates both directions, but it does allow you to configure rules in the forward chain that the older implementation did not. We'll see the difference that makes in some examples a little later.

The commands are much the same as firewall rules, except that the policy rules do not apply here. We can add, insert, delete, and list accounting rules. In the case of ipchains and iptables, all valid rules are accounting rules, and any command that doesn't specify the *-j* option performs accounting only.

The rule specification parameters for IP accounting are the same as those used for IP firewall. These are what we use to define precisely what network traffic we wish to count and total.

7.6.3.1 Accounting by Address

Let's work with an example to illustrate how we'd use IP accounting.

Imagine we have a Linux-based router that serves two departments at the Virtual Brewery. The router has two Ethernet devices, *eth0* and *eth1*, each of which services a department; and a PPP device, *ppp0*, that connects us via a high-speed serial link to the main campus of the Groucho Marx University.

Let's also imagine that for billing purposes we want to know the total traffic generated by each of the departments across the serial link, and for management purposes we want to know the total traffic generated between the two departments.

The following table shows the interface addresses we will use in our example:

iface	address	netmask
eth0	172.16.3.0	255.255.255.0
eth1	172.16.4.0	255.255.255.0

To answer the question, "How much data does each department generate on the PPP link?", we could use a rule that looks like this:

```
# ipfwadm -A both -a -W ppp0 -S 172.16.3.0/24 -b
# ipfwadm -A both -a -W ppp0 -S 172.16.4.0/24 -b
```

or:

```
# ipchains -A input -i ppp0 -d 172.16.3.0/24
# ipchains -A output -i ppp0 -s 172.16.3.0/24
# ipchains -A input -i ppp0 -d 172.16.4.0/24
# ipchains -A output -i ppp0 -s 172.16.4.0/24
```

and with iptables:

```
# iptables -A FORWARD -i ppp0 -d 172.16.3.0/24
# iptables -A FORWARD -o ppp0 -s 172.16.3.0/24
# iptables -A FORWARD -i ppp0 -d 172.16.4.0/24
# iptables -A FORWARD -o ppp0 -s 172.16.4.0/24
```

The first half of each of these set of rules say, "Count all data traveling in either direction across the interface named ppp0 with a source or destination (remember the function of the *-b* flag in ipfwadm and iptables) address of 172.16.3.0/24." The second half of each ruleset is the same, but for the second Ethernet network at our site.

To answer the second question, "How much data travels between the two departments?", we need a rule that looks like this:

```
# ipfwadm -A both -a -S 172.16.3.0/24 -D 172.16.4.0/24 -b
```

or:

```
# ipchains -A forward -s 172.16.3.0/24 -d 172.16.4.0/24 -b
```

or:

```
# iptables -A FORWARD -s 172.16.3.0/24 -d 172.16.4.0/24
# iptables -A FORWARD -s 172.16.4.0/24 -d 172.16.3.0/24
```

These rules will count all datagrams with a source address belonging to one of the department networks and a destination address belonging to the other.

7.6.3.2 Accounting by Service Port

Okay, let's suppose we also want a better idea of exactly what sort of traffic is being carried across our PPP link. We might, for example, want to know how much of the link the FTP, smtp, and World Wide Web services are consuming.

A script of rules to enable us to collect this information might look like:

```
#!/bin/sh
# Collect FTP, smtp and www volume statistics for data carried on our
# PPP link using ipfwadm
#
ipfwadm -A both -a -W ppp0 -P tcp -S 0/0 ftp ftp-data
ipfwadm -A both -a -W ppp0 -P tcp -S 0/0 smtp
ipfwadm -A both -a -W ppp0 -P tcp -S 0/0 www
```

or:

```
#!/bin/sh
# Collect ftp, smtp and www volume statistics for data carried on our
# PPP link using ipchains
#
ipchains -A input -i ppp0 -p tcp -s 0/0 ftp-data:ftp
ipchains -A output -i ppp0 -p tcp -d 0/0 ftp-data:ftp
```

```
ipchains -A input -i ppp0 -p tcp -s 0/0 smtp
ipchains -A output -i ppp0 -p tcp -d 0/0 smtp
ipchains -A input -i ppp0 -p tcp -s 0/0 www
ipchains -A output -i ppp0 -p tcp -d 0/0 www
```

or:

```
#!/bin/sh
# Collect ftp, smtp and www volume statistics for data carried on our
# PPP link using iptables.
#
iptables -A FORWARD -i ppp0 -m tcp -p tcp –sport ftp-data:ftp
iptables -A FORWARD -o ppp0 -m tcp -p tcp –dport ftp-data:ftp
iptables -A FORWARD -i ppp0 -m tcp -p tcp –sport smtp
iptables -A FORWARD -o ppp0 -m tcp -p tcp –dport smtp
iptables -A FORWARD -i ppp0 -m tcp -p tcp –sport www
iptables -A FORWARD -o ppp0 -m tcp -p tcp –dport www
```

There are a couple of interesting features to this configuration. Firstly, we've specified the protocol. When we specify ports in our rules, we must also specify a protocol because TCP and UDP provide separate sets of ports. Since all of these services are TCB-based, we've specified it as the protocol. Secondly, we've specified the two services ftp and ftp-data in one command. ipfwadm allows you to specify single ports, ranges of ports, or arbitrary lists of ports. The ipchains command allows either single ports or ranges of ports, which is what we've used here. The syntax "ftp-data:ftp" means "ports ftp-data (20) through ftp (21)," and is how we encode ranges of ports in both ipchains and iptables. When you have a list of ports in an accounting rule, it means that any data received for any of the ports in the list will cause the data to be added to that entry's totals. Remembering that the FTP service uses two ports, the command port and the data transfer port, we've added them together to total the FTP traffic. Lastly, we've specified the source address as "0/0," which is special notation that matches all addresses and is required by both the ipfwadm and ipchains commands in order to specify ports.

We can expand on the second point a little to give us a different view of the data on our link. Let's now imagine that we class FTP, SMTP, and World Wide Web traffic as essential traffic, and all other traffic as nonessential. If we were interested in seeing the ratio of essential traffic to nonessential traffic, we could do something like:

```
# ipfwadm -A both -a -W ppp0 -P tcp -S 0/0 ftp ftp-data smtp www
# ipfwadm -A both -a -W ppp0 -P tcp -S 0/0 1:19 22:24 26:79 81:32767
```

If you have already examined your */etc/services* file, you will see that the second rule covers all ports except (ftp, ftp-data, smtp, and www).

How do we do this with the ipchains or iptables commands, since they allow only one argument in their port specification? We can exploit user-defined chains in accounting just as easily as in firewall rules. Consider the following approach:

```
# ipchains -N a-essent
# ipchains -N a-noness
# ipchains -A a-essent -j ACCEPT
# ipchains -A a-noness -j ACCEPT
# ipchains -A forward -i ppp0 -p tcp -s 0/0 ftp-data:ftp -j a-essent
```

```
# ipchains -A forward -i ppp0 -p tcp -s 0/0 smtp -j a-essent
# ipchains -A forward -i ppp0 -p tcp -s 0/0 www -j a-essent
# ipchains -A forward -j a-noness
```

Here we create two user-defined chains, one called a-essent, where we capture accounting data for essential services and another called a-noness, where we capture accounting data for nonessential services. We then add rules to our forward chain that match our essential services and jump to the a-essent chain, where we have just one rule that accepts all datagrams and counts them. The last rule in our forward chain is a rule that jumps to our a-noness chain, where again we have just one rule that accepts all datagrams and counts them. The rule that jumps to the a-noness chain will not be reached by any of our essential services, as they will have been accepted in their own chain. Our tallies for essential and nonessential services will therefore be available in the rules within those chains. This is just one approach you could take; there are others. Our iptables implementation of the same approach would look like:

```
# iptables -N a-essent
# iptables -N a-noness
# iptables -A a-essent -j ACCEPT
# iptables -A a-noness -j ACCEPT
# iptables -A FORWARD -i ppp0 -m tcp -p tcp –sport ftp-data:ftp -
j a-essent
# iptables -A FORWARD -i ppp0 -m tcp -p tcp –sport smtp -j a-essent
# iptables -A FORWARD -i ppp0 -m tcp -p tcp –sport www -j a-essent
# iptables -A FORWARD -j a-noness
```

This looks simple enough. Unfortunately, there is a small but unavoidable problem when trying to do accounting by service type. You will remember that we discussed the role the MTU plays in TCP/IP networking in an earlier chapter. The MTU defines the largest datagram that will be transmitted on a network device. When a datagram is received by a router that is larger than the MTU of the interface that needs to retransmit it, the router performs a trick called *fragmentation*. The router breaks the large datagram into small pieces no longer, than the MTU of the interface and then transmits these pieces. The router builds new headers to put in front of each of these pieces, and these are what the remote machine uses to reconstruct the original data. Unfortunately, during the fragmentation process the port is lost for all but the first fragment. This means that the IP accounting can't properly count fragmented datagrams. It can reliably count only the first fragment, or unfragmented datagrams. There is a small trick permitted by ipfwadm that ensures that while we won't be able to know exactly what port the second and later fragments were from, we can still count them. An early version of Linux accounting software assigned the fragments a fake port number, 0xFFFF, that we could count. To ensure that we capture the second and later fragments, we could use a rule like:

```
# ipfwadm -A both -a -W ppp0 -P tcp -S 0/0 0xFFFF
```

The IP chains implementation has a slightly more sophisticated solution, but the result is much the same. If using the ipchains command we'd instead use:

```
# ipchains -A forward -i ppp0 -p tcp -f
```

and with iptables we'd use:

```
# iptables -A FORWARD -i ppp0 -m tcp -p tcp -f
```

These won't tell us what the original port for this data was, but at least we are able to see how much of our data is fragments, and be able to account for the volume of traffic they consume.

In 2.2 kernels you can select a kernel compile-time option that negates this whole issue if your Linux machine is acting as the single access point for a network. If you enable the *IP: always defragment* option when you compile your kernel, all received datagrams will be reassembled by the Linux router before routing and retransmission. This operation is performed before the firewall and accounting software sees the datagram, and thus you will have no fragments to deal with. In 2.4 kernels you compile and load the *netfilter forward-fragment* module.

7.6.3.3 Accounting of ICMP Datagrams

The ICMP protocol does not use service port numbers and is therefore a little bit more difficult to collect details on. ICMP uses a number of different types of datagrams. Many of these are harmless and normal, while others should only be seen under special circumstances. Sometimes people with too much time on their hands attempt to maliciously disrupt the network access of a user by generating large numbers of ICMP messages. This is commonly called *ping flooding.* While IP accounting cannot do anything to prevent this problem (IP firewalling can help, though!) we can at least put accounting rules in place that will show us if anybody has been trying.

ICMP doesn't use ports as TCP and UDP do. Instead ICMP has ICMP message types. We can build rules to account for each ICMP message type. To do this, we place the ICMP message and type number in place of the port field in the ipfwadm accounting commands. We listed the ICMP message types in "ICMP datagram types", so refer to it if you need to remember what they are.

An IP accounting rule to collect information about the volume of ping data that is being sent to you or that you are generating might look like:

```
# ipfwadm -A both -a -P icmp -S 0/0 8
# ipfwadm -A both -a -P icmp -S 0/0 0
# ipfwadm -A both -a -P icmp -S 0/0 0xff
```

or, with `ipchains`:

```
# ipchains -A forward -p icmp -s 0/0 8
# ipchains -A forward -p icmp -s 0/0 0
# ipchains -A forward -p icmp -s 0/0 -f
```

or, with `iptables`:

```
# iptables -A FORWARD -m icmp -p icmp —sports echo-request
# iptables -A FORWARD -m icmp -p icmp —sports echo-reply
# iptables -A FORWARD -m icmp -p icmp -f
```

The first rule collects information about the "ICMP Echo Request" datagrams (ping requests), and the second rule collects information about the "ICMP Echo Reply" datagrams (ping replies). The third rule collects information about ICMP datagram fragments. This is a trick similar to that described for fragmented TCP and UDP datagrams.

If you specify source and/or destination addresses in your rules, you can keep track of where the pings are coming from, such as whether they originate inside or outside your network. Once you've determined where the rogue datagrams are coming from, you can decide whether you want to put firewall rules in place to prevent them or take some other action, such as contacting the owner of the offending network to advise them of the problem, or perhaps even legal action if the problem is a malicious act.

7.6.3.4 Accounting by Protocol

Let's now imagine that we are interested in knowing how much of the traffic on our link is TCP, UDP, and ICMP. We would use rules like the following:

```
# ipfwadm -A both -a -W ppp0 -P tcp -D 0/0
# ipfwadm -A both -a -W ppp0 -P udp -D 0/0
# ipfwadm -A both -a -W ppp0 -P icmp -D 0/0
```

or:

```
# ipchains -A forward -i ppp0 -p tcp -d 0/0
# ipchains -A forward -i ppp0 -p udp -d 0/0
# ipchains -A forward -i ppp0 -p icmp -d 0/0
```

or:

```
# iptables -A FORWARD -i ppp0 -m tcp -p tcp
# iptables -A FORWARD -o ppp0 -m tcp -p tcp
# iptables -A FORWARD -i ppp0 -m udp -p udp
# iptables -A FORWARD -o ppp0 -m udp -p udp
# iptables -A FORWARD -i ppp0 -m icmp -p icmp
# iptables -A FORWARD -o ppp0 -m icmp -p icmp
```

With these rules in place, all of the traffic flowing across the ppp0 interface will be analyzed to determine whether it is TCP, UDP, or IMCP traffic, and the appropriate counters will be updated for each. The iptables example splits incoming flow from outgoing flow as its syntax demands it.

7.6.4 Using IP Accounting Results

It is all very well to be collecting this information, but how do we actually get to see it? To view the collected accounting data and the configured accounting rules, we use our firewall configuration commands, asking them to list our rules. The packet and byte counters for each of our rules are listed in the output.

The ipfwadm, ipchains, and iptables commands differ in how accounting data is handled, so we will treat them independently.

7.6.4.1 Listing Accounting Data with ipfwadm

The most basic means of listing our accounting data with the ipfwadm command is to use it like this:

```
# ipfwadm -A -l
IP accounting rules
pkts bytes dir prot source destination ports
9833 2345K i/o all 172.16.3.0/24 anywhere n/a
56527 33M i/o all 172.16.4.0/24 anywhere n/a
```

This will tell us the number of packets sent in each direction. If we use the extended output format with the *-e* option (not shown here because the output is too wide for the page), we are also supplied the options and applicable interface names. Most of the fields in the output will be self-explanatory, but the following may not:

dir

The direction in which the rule applies. Expected values here are in, out, or i/o, meaning both ways.

prot

The protocols to which the rule applies.

opt

A coded form of the options we use when invoking ipfwadm.

ifname

The name of the interface to which the rule applies.

ifaddress

The address of the interface to which the rule applies.

By default, ipfwadm displays the packet and byte counts in a shortened form, rounded to the nearest thousand (K) or million (M). We can ask it to display the collected data in exact units by using the expanded option as follows:

```
# ipfwadm -A -l -e -x
```

7.6.4.2 Listing Accounting Data with ipchains

The ipchains command will not display our accounting data (packet and byte counters) unless we supply it the -v argument. The simplest means of listing our accounting data with the ipchains is to use it like this:

```
# ipchains -L -v
```

Again, just as with ipfwadm, we can display the packet and byte counters in units by using the expanded output mode. The ipchains uses the -x argument for this:

```
# ipchains -L -v -x
```

7.6.4.3 Listing Accounting Data with iptables

The iptables command behaves very similarly to the ipchains command. Again, we must use the -v when listing tour rules to see the accounting counters. To list our accounting data, we would use:

```
# iptables -L -v
```

Just as for the ipchains command, you can use the -x argument to show the output in expanded format with unit figures.

7.7 IP MASQUERADE AND NETWORK ADDRESS TRANSLATION

You don't have to have a good memory to remember a time when only large organizations could afford to have a number of computers networked together by a LAN. Today network technology has dropped so much in price that two things have happened. First, LANs are now commonplace, even in many household environments. Certainly many Linux users will have two or more computers connected by some Ethernet. Second, network resources, particularly IP addresses, are now a scarce resource and while they used to be free, they are now being bought and sold.

Most people with a LAN will probably also want an Internet connection that every computer on the LAN can use. The IP routing rules are quite strict in how they deal with this situation. Traditional solutions to this problem would have involved requesting an IP network address, perhaps a class C address for small sites, assigning each host on the LAN an address from this network and using a router to connect the LAN to the Internet.

In a commercialized Internet environment, this is quite an expensive proposition. First, you'd be required to pay for the network address that is assigned to you. Second, you'd probably have to pay your Internet Service Provider for the privilege of having a suitable route to your network put in place so that the rest of the Internet knows how to reach you. This might still be practical for companies, but domestic installations don't usually justify the cost.

Fortunately, Linux provides an answer to this dilemma. This answer involves a component of a group of advanced networking features called *Network Address Translation* (NAT). NAT describes the process of modifying the network addresses contained with datagram headers while they are in transit. This might sound odd at first, but we'll show that it is ideal for solving the problem we've just described and many have encountered. IP masquerade is the name given to one type of network address translation that allows all of the hosts on a private network to use the Internet at the price of a single IP address.

IP masquerading allows you to use a private (reserved) IP network address on your LAN and have your Linux-based router perform some clever, real-time translation of IP addresses and ports. When it receives a datagram from a computer on the LAN, it takes note of the type of datagram it is, "TCP," "UDP," "ICMP," etc., and modifies the datagram so that it looks like it was generated by the router machine itself (and remembers that it has done so). It then transmits the datagram onto the Internet with its single connected IP address. When the destination host receives this datagram, it believes the datagram has come from the routing host and sends any reply datagrams back to that address. When the Linux masquerade router receives a datagram from its Internet connection, it looks in its table of established masqueraded connections to see if this datagram actually belongs to a computer on the LAN, and if it does, it reverses the modification it did on the forward path and transmits the datagram to the LAN computer.

7.8 SIDE EFFECTS AND FRINGE BENEFITS

The IP masquerade facility comes with its own set of side effects, some of which are useful and some of which might become bothersome.

None of the hosts on the supported network behind the masquerade router are ever directly seen; consequently, you need only one valid and routable IP address to allow all hosts to make network connections out onto the Internet. This has a downside; none of those hosts are visible from the Internet and you can't directly connect to them from the Internet; the only host visible on a masqueraded network is the masquerade machine itself. This is important when you consider services such as mail or FTP. It helps determine what services should be provided by the masquerade host and what services it should proxy or otherwise treat specially.

Second, because none of the masqueraded hosts are visible, they are relatively protected from attacks from outside; this could simplify or even remove the need for firewall configuration on the masquerade host. You shouldn't rely too heavily on this, though. Your whole network will be only as safe as your masquerade host, so you should use firewall to protect it if security is a concern.

Third, IP masquerade will have some impact on the performance of your networking. In typical configurations this will probably be barely measurable. If you have large numbers of active masquerade sessions, though, you may find that the processing required at the masquerade machine begins to impact your network throughput. IP masquerade must do a good deal of work for each datagram compared to the process of conventional routing. That 386SX16 machine you have been planning on using as a masquerade machine supporting a dial-up link to the Internet might be fine, but don't expect too much if you decide you want to use it as a router in your corporate network at Ethernet speeds.

Last, some network services just won't work through masquerade, or at least not without a lot of help. Typically, these are services that rely on incoming sessions to work, such as some types of Direct Communications Channels (DCC), features in IRC, or certain types of video and audio multicasting services. Some of these services have specially developed kernel modules to provide solutions for these, and we'll talk about those in a moment. For others, it is possible that you will find no support, so be aware,it won't be suitable in all situations.

7.9 CONFIGURING THE KERNEL FOR IP MASQUERADE

To use the IP masquerade facility, your kernel must be compiled with masquerade support. You must select the following options when configuring a 2.2 series kernel:

```
Networking options ->
    [*] Network firewalls
    [*] TCP/IP networking
    [*] IP: firewalling
    [*] IP: masquerading
    – Protocol-specific masquerading support will be built as modules.
    [*] IP: ipautofw masq support
    [*] IP: ICMP masquerading
```

Note that some of the masquerade support is available only as a kernel module. This means that you must ensure that you "make modules" in addition to the usual "make zImage" when building your kernel.

The 2.4 series kernels no longer offer IP masquerade support as a kernel compile time option. Instead, you should select the network packet filtering option:

```
Networking options ->
    [M] Network packet filtering (replaces ipchains)
```

In the 2.2 series kernels, a number of protocol-specific helper modules are created during kernel compilation. Some protocols begin with an outgoing request on one port, and then expect an incoming connection on another. Normally, these cannot be masqueraded, as there is no way of associating the second connection with the first without peering inside the protocols themselves. The helper modules do just that; they actually look inside the datagrams and allow masquerading to work for supported protocols that otherwise would be impossible to masquerade. The supported protocols are:

Module	Protocol
ip_masq_ftp	FTP
ip_masq_irc	IRC
ip_masq_raudio	RealAudio
ip_masq_cuseeme	CU-See-Me

Module	Protocol
ip_masq_vdolive	For VDO Live
ip_masq_quake	IdSoftware's Quake

You must load these modules manually using the insmod command to implement them. Note that these modules cannot be loaded using the kerneld daemon. Each of the modules takes an argument specifying what ports it will listen on. For the RealAudio(TM) module you might use

```
# insmod ip_masq_raudio.o ports=7070,7071,7072
```

The ports you need to specify depend on the protocol. An IP masquerade mini-HOWTO written by Ambrose Au explains more about the IP masquerade modules and how to configure them.[2]

RealAudio is a trademark of the Progressive Networks Corporation.

The *netfilter* package includes modules that perform similar functions. For example, to provide connection tracking of FTP sessions, you'd load and use the *ip_conntrack_ftp* and *ip_nat_ftp.o* modules.

7.10 CONFIGURING IP MASQUERADE

If you've already read the firewall and accounting chapters, it probably comes as no surprise that the ipfwadm, ipchains, and iptables commands are used to configure the IP masquerade rules as well.

Masquerade rules are a special class of filtering rule. You can masquerade only datagrams that are received on one interface that will be routed to another interface. To configure a masquerade rule you construct a rule very similar to a firewall forwarding rule, but with special options that tell the kernel to masquerade the datagram. The ipfwadm command uses the *-m* option, ipchains uses -j MASQ, and iptables uses -j MASQUERADE to indicate that datagrams matching the rule specification should be masqueraded.

Let's look at an example. A computing science student at Groucho Marx University has a number of computers at home internetworked onto a small Ethernet-based local area network. She has chosen to use one of the reserved private Internet network addresses for her network. She shares her accomodation with other students, all of whom have an interest in using the Internet. Because student living conditions are very frugal, they cannot afford to use a permanent Internet connection, so instead they use a simple dial-up PPP Internet connection. They would all like to be able to share the connection to chat on IRC, surf the Web, and retrieve files by FTP directly to each of their computers — IP masquerade is the answer.

The student first configures a Linux machine to support the dial-up link and to act as a router for the LAN. The IP address she is assigned when she dials up isn't important. She configures the Linux router with IP masquerade and uses one of the private network addresses for her LAN: 192.168.1.0. She ensures that each of the hosts on the LAN has a default route pointing at the Linux router.

The following ipfwadm commands are all that are required to make masquerading work in her configuration:

```
# ipfwadm -F -p deny
# ipfwadm -F -a accept -m -S 192.168.1.0/24 -D 0/0
```

or with `ipchains`:

```
# ipchains -P forward -j deny
# ipchains -A forward -s 192.168.1.0/24 -d 0/0 -j MASQ
```

or with iptables:

```
# iptables -t nat -P POSTROUTING DROP
# iptables -t nat -A POSTROUTING -o ppp0 -j MASQUERADE
```

Now whenever any of the LAN hosts try to connect to a service on a remote host, their datagrams will be automatically masqueraded by the Linux masquerade router. The first rule in each example prevents the Linux machine from routing any other datagrams and also adds some security.

To list the masquerade rules you have created, use the -l argument to the ipfwadm command, as we described in earlier while discussing firewalls.

To list the rule we created earlier we use:

```
# ipfwadm -F -l -e
```

which should display something like:

```
# ipfwadm -F -l -e
IP firewall forward rules, default policy: accept
 pkts bytes type prot opt tosa tosx ifname ifaddress ...
 0 0 acc/m all -- 0xFF 0x00 any any ...
```

The "/m" in the output indicates this is a masquerade rule.

To list the masquerade rules with the ipchains command, use the -L argument. If we list the rule we created earlier with ipchains, the output will look like:

```
# ipchains -L
Chain input (policy ACCEPT):
Chain forward (policy ACCEPT):
target prot opt source destination ports
MASQ all --- 192.168.1.0/24 anywhere n/a
Chain output (policy ACCEPT):
```

Any rules with a target of MASQ are masquerade rules.

Finally, to list the rules using iptables you need to use:

```
# iptables -t nat -L
Chain PREROUTING (policy ACCEPT)
target prot opt source destination

Chain POSTROUTING (policy DROP)
target prot opt source destination
MASQUERADE all - anywhere anywhere MASQUERADE

Chain OUTPUT (policy ACCEPT)
target prot opt source destination
```

Again, masquerade rules appear with a target of MASQUERADE.

7.10.1 Setting Timing Parameters for IP Masquerade

When each new connection is established, the IP masquerade software creates an association in memory between each of the hosts involved in the connection. You can view these associations at any time by looking at the */proc/net/ip_masquerade* file. These associations will timeout after a period of inactivity, though.

You can set the timeout values using the ipfwadm command. The general syntax for this is:

```
ipfwadm -M -s <tcp> <tcpfin> <udp>
```

and for the ipchains command it is:

```
ipchains -M -S <tcp> <tcpfin> <udp>
```

The *iptables* implementation uses much longer default timers and does not allow you to set them.

Each of these values represents a timer used by the IP masquerade software and are in units of seconds. The following table summarizes the timers and their meanings:

Name	Description
tcp	TCP session timeout. How long a TCP connection may remain idle before the association for it is removed.
tcpfin	TCP timeout after FIN. How long an association will remain after a TCP connection has been disconnected.
udp	UDP session timeout. How long a UDP connection may remain idle before the association for it is removed.

SUMMARY

1. Security is increasingly important for companies and individuals alike. The Internet has provided them with a powerful tool to distribute information about themselves and obtain information from others, but it has also exposed them to dangers that they have previously been exempt from.
2. Computer crime, information theft, and malicious damage are all potential dangers. An unauthorized and unscrupulous person who gains access to a computer system may guess system passwords or exploit the bugs and idiosyncratic behavior of certain programs to obtain a working account on that machine.
3. Once they are able to log in to the machine, they may have access to information that may be damaging, such as commercially sensitive information like marketing plans, new project details, or customer information databases.
4. Damaging or modifying this type of data can cause severe setbacks to the company.
5. The safest way to avoid such widespread damage is to prevent unauthorized people from gaining network access to the machine. This is where firewalls come in.
6. Denial of service attacks cause the service or program to cease functioning or prevent others from making use of the service or program. These may be performed at the network layer by sending carefully crafted and malicious datagrams that cause network connections to fail.
7. They may also be performed at the application layer, where carefully crafted application commands are given to a program that cause it to become extremely busy or stop functioning.
8. Preventing suspicious network traffic from reaching your hosts and preventing suspicious program commands and requests are the best ways of minimizing the risk of a denial of service attack. It's useful to know the details of the attack method, so you should educate yourself about each new attack as it gets publicized.

9. Typically the attacker pretends to be an innocent host by following IP addresses in network packets. For example, a well-documented exploit of the BSD rlogin service can use this method to mimic a TCP connection from another host by guessing TCP sequence numbers.
10. To protect against this type of attack, verify the authenticity of datagrams and commands. Prevent datagram routing with invalid source addresses. Introduce unpredictablility into connection control mechanisms, such as TCP sequence numbers and the allocation of dynamic port addresses.

SHORT QUESTIONS WITH ANSWERS

1. Change /etc/inittab so that only two mingetty's are respawned. Kill the other mingetty's and verify that they don't come back.
Killing the mingetty's will result in init respawning them. You can edit /etc/inittab so it looks like the screenshot below. Don't forget to also run kill -1 1.

Ans.
```
[root@RHEL5 -]# grep tty /etc/inittab
# Run gettys in standard runlevels
1:2345:respawn:/sbin/mingetty tty1
2:2345:respawn:/sbin/mingety tty2
3:2:respawn:/sbin/mingetty tty3
4:2:respawn:/sbin/mingetty tty4
5:2:respawn:/sbin/mingetty tty5
6:2:respawn:/sbin/mingetty tty6
[root@RHEL5 ~]#
```

2. Use the Red Hat Enterprise Linux virtual machine. Go to nmlevel 5, display the current and previous nmlevel, then go back to nmlevel 3.

Ans.
```
init 5 (watch the console for the change taking place)
runlevel
init 3 (again you can follow this on the console)
```

3. Is the sysinit script on your computers setting or changing the PATH environment variable ?

Ans. On Red Hat, grep for PATH in **/etc/rc.sysinit**, on Debian/Ubuntu check **/etc/rc.local** and **/etc/ini.t/rc.local**. The answer is probably no, but on RHEL5 the **rc.sysinit** script does set the HOSTNAME variable.

Ans.
```
[rootaRHEL5 etc]# grep HOSTNAME rc.sysinit
```

4. List all init.d scripts that are started in runlevel 2.

Ans.
```
root@RHEL5 ~# chkconfig --list | grep '2:on'
```

5. Write a script mat acts like a daemon script in /etc/init.d/. It should have a case statement to act on start/stop/restart and status. Test the script!

Ans. The script could look something like this.
```
#!/bin/bash
#
# chkconfig: 345 99 01
# description: pold demo script
#
# /etc/init.d/pold
#
case "$1" in
 start)
     echo -n "Starting pold..."
```

```
        sleep 1;
        touch /var/lock/subsys/pold
        echo "done."
        echo pold started >> /var/log/messages
        ; ;
    stop)
        echo -n "Stopping pold..."
        sleep 1;
        rm -rf /var/lock/subsys/pold
        echo "done."
        echo pold stopped >> /var/log/messages
        ; ;
    *)
        echo "Usage: /etc/init.d/pold (start|stop}"
        exit 1
    esac
    exit 0
```

The **touch /var/lock/subsys/pold** is mandatory and must be the same filename as the script name, if you want the stop sequence (the K01pold link) to be run.

6. Use chkconfig to setap your script to start in mnlevels 3,4 and 5, and to stop in any other nmlevel.

Ans. chkconfig --add pold

7. Schedule two jobs with at, display the at queue and remove a job.

Ans.
```
root@rhel55 -tf at 9pm today
at> echo go to bed >> /root/todo.txt
at> <BOT>
job 1 at 2010-11-14 21:00
root@rhel55 ~# at 17h31 today
at> echo go to lunch >> /root/todo.txt
at> <EOT>
job 2 at 2010-11-14 17:31
root@rhel55 -# atq
2 2010-11-14 17:31 a root
1 2010-11-14 21:00 a root
root@rhel55 ~# atrm 1
root@rhel55~# atq
2 2010-11-14 17:31 a root
root@rhel55 -# date
Sun Nov 14 17:31:01 CET 2010
root@rhel55 -# cat /root/todo.txt
go to lunch
```

8. As normal user, use crontab -e to schedule a script to run every four minutes.

Ans.
```
paul@rhel55 ~$ crontab -e
no crontab for paul - using an empty one
crontab: installing new crontab
```

9. As root, display the crontab file of your normal user.

Ans.
```
root@rehel55 ~# crontab -1 -u paul
*/4 * * * * echo ~date~ >> /home/paul/crontest.txt
```

10. As the normal user again, remove your crontab file.

Ans.
```
paul@rhel55 ~$ crontab -r
paul@rhel55 -$ crontab -1
no crontab for paul
```

11. Take a look at the cron files and directories in /etc and understand them. What is the run-parts command doing ?

Ans. `run-parts runs a script in a directory.`

EXERCISE

1. What is a process?
2. Can a series of process be run together?
3. How does UNIX Identify a process?
4. How do you send a process in a Background.
5. Can you kill a process running in Background & How.
6. Write a Shell script to check if the input string is a palindrome.
7. Write a Shell script to accept a number and a word as command line arguments and print the word the given number of times on each line. What happens if a number and a string is given as command line arguments ?
8. Write a Shell script to find the file with the maximum size in the current directory.
9. Write a Shell script to accept an input and check if the given input is a directory. If it is a directory, then display its contents and revoke the execute permission for group and others for all files starting with "b" in the directory.
10. Write a Shell script to accept two filenames and check if both exist. If the second filename exists, then the contents of the first filename should be appended to it. If the second filename does not exist then create a newfile with the contents of the first file.
11. Write a Shell script to accept a number in the command line and displays the sum up to that number. For example, if the number given is 20, then the sum of numbers 1,2,3,...,20 should be displayed. By default, the sum upto 50 should be displayed.
12. Write a Shell script to find the number of ordinary files and directory files in the current directory.
13. Write a Shell script to accept an alphabet from the user and list all the files starting with that letter starting from the current directory.
14. Write a Shell script to accept the name of the directory as command line argument and display the long listing of that directory. By default the HOME directory's contents should be displayed.
15. Write a Shell script which accepts a number and displays the list of odd numbers below that number.

❒❒❒

CHAPTER 8

THE NETWORK INFORMATION SYSTEM

8.0 INTRODUCTION

Chapter Outline

- Introduction
- The Network Information System
- Getting Acquainted with NIS
- NIS Versus NIS+
- The Client Side of NIS
- Running an NIS Server
- NIS Server Security
- Setting Up an NIS Client with GNU LIBC
- Network File System
- Preparing NFS
- Mounting an NFS Volume
- The NFS Daemons
- The Exports File
- System Backup & Recovery
- Log Files for System and Applications
- Backup Schedules and Methods (Manual and Automated)

When you're running a local area network, your overall goal is usually to provide an environment for your users that makes the network transparent. An important stepping stone is keeping vital data such as user account information synchronized among all hosts. This provides users with the freedom to move from machine-to-machine without the inconvenience of having to remember different passwords and copy data from one machine to another. Data that is centrally stored doesn't need to be replicated, so long as there is some convenient means of accessing it from a network-connected host. By storing important administrative information centrally, you can make ensure consistency of that data, increase flexibility for the users by allowing them to move from host-to-host in a transparent way, and make the system administrator's life much easier by maintaining a single copy of information to maintain when required.

We previously discussed an important example of this concept that is used on the Internet — the Domain Name System (DNS). DNS serves a limited range of information, the most important being the mapping between hostname and IP address. For other types of information, there is no such specialized service. Moreover, if you manage only a small LAN with no Internet connectivity, setting up DNS may not seem to be worth the trouble.

8.1 THE NETWORK INFORMATION SYSTEM

This is why Sun developed the *Network Information System* (NIS). NIS provides generic database access facilities that can be used to distribute, for example, information contained in the *passwd* and *groups* files to all hosts on your network. This makes the network appear as a single system, with the same accounts on all hosts. Similarly, you can use NIS to distribute the hostname information from */etc/hosts* to all machines on the network.

NIS is based on RPC, and comprises a server, a client-side library, and several administrative tools. Originally, NIS was called *Yellow Pages*, or YP, which is still used to refer to it. Unfortunately, the name is a trademark of British Telecom, which required Sun to drop that name. As things go, some names stick with people, and so YP lives on as a prefix to the names of most NIS-related commands such as ypserv and ypbind.

Today, NIS is available for virtually all Unixes, and there are even free implementations. BSD Net-2 released one that has been derived from a public domain reference implementation donated by Sun.

8.2 GETTING ACQUAINTED WITH NIS

NIS keeps database information in files called *maps*, which contain key-value pairs. An example of a key-value pair is a user's login name and the encrypted form of their login password. Maps are stored on a central host running the NIS server, from which clients may retrieve the information through various RPC calls. Quite frequently, maps are stored in DBM files

DBM is a simple database management library that uses hashing techniques to speed up search operations. There's a free DBM implementation from the GNU project called *gdbm*, which is part of most Linux distributions.

The maps themselves are usually generated from master text files such as */etc/hosts* or */etc/passwd*. For some files, several maps are created, one for each search key type. For instance, you may search the *hosts* file for a hostname as well as for an IP address. Accordingly, two NIS maps are derived from it, called *hosts.byname* and *hosts.byaddr*. Table 8.1 lists common maps and the files from which they are generated.

Table 8.1: Some Standard NIS Maps and Corresponding Files

Master File	Map(s)	Description
/etc/hosts	*hosts.byname, hosts.byaddr*	Maps IP addresses to host names
/etc/networks	*networks.byname, networks.byaddr*	Maps IP network addresses to network names
/etc/passwd	*passwd.byname, passwd.byuid*	Maps encrypted passwords to user login names
/etc/group	*group.byname, group.bygid*	Maps Group IDs to group names
/etc/services	*services.byname, services.bynumber*	Maps service descriptions to service names
/etc/rpc	*rpc.byname, rpc.bynumber*	Maps Sun RPC service numbers to RPC service names
/etc/protocols	*protocols.byname, protocols.bynumber*	Maps protocol numbers to protocol names
/usr/lib/aliases	*mail.aliases*	Maps mail aliases to mail alias names

You may find support for other files and maps in other NIS packages. These usually contain information for applications not discussed in this book, such as the *bootparams* map that is used by Sun's bootparamd server.

For some maps, people commonly use *nicknames*, which are shorter and therefore easier to type. Note that these nicknames are understood only by ypcat and ypmatch, two tools for checking your NIS configuration. To obtain a full list of nicknames understood by these tools, run the following command:

```
$ ypcat -x
Use "passwd" for "passwd.byname"
Use "group" for "group.byname"
Use "networks" for "networks.byaddr"
Use "hosts" for "hosts.byaddr"
Use "protocols" for "protocols.bynumber"
Use "services" for "services.byname"
Use "aliases" for "mail.aliases"
Use "ethers" for "ethers.byname"
```

The NIS server program is traditionally called ypserv. For an average network, a single server usually suffices; large networks may choose to run several of these on different machines and different segments of the network to relieve the load on the server machines and routers. These servers are synchronized by making one of them the *master server*, and the others *slave servers*. Maps are created only on the master server's host. From there, they are distributed to all slaves.

We have been talking very vaguely about "networks." There's a distinctive term in NIS that refers to a collection of all hosts that share part of their system configuration data through NIS: the *NIS domain*. Unfortunately, NIS domains have absolutely nothing in common with the domains we encountered in DNS. To avoid any ambiguity throughout this chapter, we will therefore always specify which type of domain we mean.

NIS domains have a purely administrative function. They are mostly invisible to users, except for the sharing of passwords between all machines in the domain. Therefore, the name given to an NIS domain is relevant only to the administrators. Usually, any name will do, as long as it is different from any other NIS domain name on your local network. For instance, the administrator at the Virtual Brewery may choose to create two NIS domains, one for the Brewery itself, and one for the Winery, which she names *brewery* and *winery* respectively. Another quite common scheme is to simply use the DNS domain name for NIS as well.

To set and display the NIS domain name of your host, you can use the domainname command. When invoked without any argument, it prints the current NIS domain name; to set the domain name, you must become the superuser:

```
# domainname brewery
```

NIS domains determine which NIS server an application will query. For instance, the login program on a host at the Winery should, of course, query only the Winery's NIS server (or one of them, if there are several) for a user's password information, while an application on a Brewery host should stick with the Brewery's server.

One mystery now remains to be solved: how does a client find out which server to connect to? The simplest approach would use a configuration file that names the host on which to find the server. However, this approach is rather inflexible because it doesn't allow clients to use different servers (from the same domain, of course) depending on their availability. Therefore, NIS implementations rely on a special daemon called ypbind to detect a suitable NIS server in their NIS domain. Before performing any NIS queries, an application first finds out from ypbind which server to use.

ypbind probes for servers by broadcasting to the local IP network; the first to respond is assumed to be the fastest one and is used in all subsequent NIS queries. After a certain interval has elapsed, or if the server becomes unavailable, ypbind probes for active servers again.

Dynamic binding is useful only when your network provides more than one NIS server. Dynamic binding also introduces a security problem. ypbind blindly believes whoever answers, whether it be a humble NIS server or a malicious intruder. Needless to say, this becomes especially troublesome if you manage your password databases over NIS. To guard against this, the Linux ypbind program provides you with the option of probing the local network to find the local NIS server, or configuring the NIS server hostname in a configuration fi.

8.3 NIS VERSUS NIS+

NIS and NIS+ share little more than their name and a common goal. NIS+ is structured entirely differently from NIS. Instead of a flat namespace with disjoint NIS domains, NIS+ uses a hierarchical namespace similar to that of DNS. Instead of maps, so-called *tables* are used that are made up of rows and columns, in which each row represents an object in the NIS+ database and the columns cover properties of the objects that NIS+ knows and cares about. Each table for a given NIS+ domain comprises those of its parent domains. In addition, an entry in a table may contain a link to another table. These features make it possible to structure information in many ways.

NIS+ additionally supports secure and encrypted RPC, which helps greatly to solve the security problems of NIS.

8.4 THE CLIENT SIDE OF NIS

If you are familiar with writing or porting network applications, you may notice that most of the NIS maps listed previously correspond to library functions in the C library. For instance, to obtain *passwd* information, you generally use the *getpwnam* and *getpwuid* functions, which return the account information associated with the given username or numerical user ID, respectively. Under normal circumstances, these functions perform the requested lookup on the standard file, such as */etc/passwd*.

An NIS-aware implementation of these functions, however, modifies this behavior and places an RPC call to the NIS server, which looks up the username or user ID. This happens transparently to the application. The function may treat the NIS data as though it has been appended to the original *passwd* file so both sets of information are available to the application and used, or as though it has completely replaced it so that the information in the local *passwd* is ignored and only the NIS data is used.

For traditional NIS implementations, there were certain conventions for which maps were replaced and which were appended to the original information. Some, like the *passwd* maps, required kludgy modifications of the *passwd* file which, when done incorrectly, would open up security holes.

8.5 RUNNING AN NIS SERVER

After so much theoretical techno-babble, it's time to get our hands dirty with actual configuration work. In this section, we will cover the configuration of an NIS server. If an NIS server is running on your network, you won't have to set up your own; in this case, you may safely skip this section.

Note that if you are just going to experiment with the server, make sure you don't set it up for an NIS domain name that is already in use on your network. This may disrupt the entire network service and make a lot of people very unhappy and very angry.

There are two possible NIS server configurations: master and slave. The slave configuration provides a live backup machine, should your master server fail. We will cover the configuration only for a master server here. The server documentation will explain the differences, should you wish to configure a slave server.

There are currently two NIS servers freely available for Linux: one contained in Tobias Reber's *yps* package, and the other in Peter Eriksson's *ypserv* package. It doesn't matter which one you run.

After installing the server program (ypserv) in */usr/sbin*, you should create the directory that is going to hold the map files your server is to distribute. When setting up an NIS domain for the *brewery* domain, the maps would go to */var/yp/brewery*. The server determines whether it is serving a particular NIS domain by checking if the map directory is present. If you are disabling service for some NIS domain, make sure to remove the directory as well.

Maps are usually stored in DBM files to speed up lookups. They are created from the master files using a program called makedbm (for Tobias's server) or dbmload (for Peter's server).

Transforming a master file into a form that dbmload can parse usually requires some awk or sed magic, which tends to be a little tedious to type and hard to remember. Therefore, Peter Eriksson's *ypserv* package contains a Makefile (called *ypMakefile*) that manages the conversion of the most common master files for you. You should install it as *Makefile* in your map directory and edit it to reflect the maps you want the NIS server to share. Towards the top of the file, you'll find the *all* target that lists the services ypserv offers. By default, the line looks something like this:

```
all: ethers hosts networks protocols rpc services passwd group netid
```

If you don't want to produce, for example, the *ethers.byname* and *ethers.byaddr* maps, simply remove the *ethers* prerequisite from this rule. To test your setup, you can start with just one or two maps, like the *services.** maps.

After editing the *Makefile*, while in the map directory, type **make**. This will automatically generate and install the maps. You have to make sure to update the maps whenever you change the master files, otherwise the changes will remain invisible to the network.

The section "Setting Up an NIS Client with GNU libc" will explain how to configure the NIS client code. If your setup doesn't work, you should try to find out whether requests are arriving at your server. If you specify the *—debug* command-line flag to ypserv, it prints debugging messages to the console about all incoming NIS queries and the results returned. These should give you a hint as to where the problem lies. Tobias's server doesn't have this option.

8.6 NIS SERVER SECURITY

NIS used to have a major security flaw: it left your password file readable by virtually anyone in the entire Internet, which made for quite a number of possible intruders. As long as an intruder knew your NIS domain name and the address of your server, he could simply send it a request for the *passwd.byname* map and instantly receive all your system's encrypted passwords. With a fast password-cracking program like crack and a good dictionary, guessing at least a few of your users' passwords is rarely a problem.

This is what the *securenets* option is all about. It simply restricts access to your NIS server to certain hosts, based on their IP addresses or network numbers. The latest version of ypserv implements this feature in two ways. Thus, to restrict access to hosts from within the Brewery, their network manager would add the following line to *hosts.allow*:

```
ypserv:  172.16.2.
```

To enable use of the */etc/hosts.allow* method, you may have to recompile the server. Please read the instructions in the *README* included in the distribution.

This would let all hosts from IP network *172.16.2.0* access the NIS server. To shut out all other hosts, a corresponding entry in *hosts.deny* would have to read:

```
ypserv:  ALL
```

IP numbers are not the only way you can specify hosts or networks in *hosts.allow* and *hosts.deny*. Please refer to the *hosts_access(5)* manual page on your system for details. However, be warned that you *cannot* use host or domain names for the *ypserv* entry. If you specify a hostname, the server tries to resolve this hostname — but the resolver in turn calls ypserv, and you fall into an endless loop.

To configure *securenets* security using the */etc/ypserv.securenets* method, you need to create its configuration file, */etc/ypserv.securenets*. This configuration file is simple in structure. Each line describes a host or network of hosts that will be allowed access to the server. Any address not described by an entry in this file will be refused access. A line beginning with a # will be treated as a comment. Example 13-1 shows what a simple /etc/ypserv.securenets would look like:

EXAMPLE 8.1 *Sample ypserv.securenets File*

```
# allow connections from local host - necessary
host 127.0.0.1
# same as 255.255.255.255 127.0.0.1
#
# allow connections from any host on the Virtual Brewery network
255.255.255.0 172.16.1.0
#
```

The first entry on each line is the netmask to use for the entry, with *host* being treated as a special keyword meaning "netmask 255.255.255.255." The second entry on each line is the IP address to which to apply the netmask.

A third option is to use the secure portmapper instead of the *securenets* option in ypserv. The secure portmapper (*portmap-5.0*) uses the *hosts.allow* scheme as well, but offers this for all RPC servers, not just ypserv. However, you should not use both the *securenets* option and the secure portmapper at the same time, because of the overhead this authorization incurs.

The secure portmapper is available via anonymous FTP from *ftp.win.tue.nl* below the */pub/security/* directory.

8.7 SETTING UP AN NIS CLIENT WITH GNU LIBC

We will now describe and discuss the configuration of an NIS client using the GNU libc library support.

Your first step should be to tell the GNU libc NIS client which server to use for NIS service. We mentioned earlier that the Linux ypbind allows you to configure the NIS server to use. The default behavior is to query the server on the local network. If the host you are configuring is likely to move from one domain to another, such as a laptop, you would leave the */etc/yp.conf* file empty and it would query on the local network for the local NIS server wherever it happens to be.

A more secure configuration for most hosts is to set the server name in the */etc/yp.conf* configuration file. A very simple file for a host on the Winery's network may look like this:

```
# yp.conf - YP configuration for GNU libc library.
#
ypserver vbardolino
```

The *ypserver* statement tells your host to use the host supplied as the NIS server for the local domain. In this example we've specified the NIS server as *vbardolino*. Of course, the IP address corresponding to *vbardolino* must be set in the *hosts* file; alternatively, you may use the IP address itself with the *server* argument.

In the form shown in the example, the ypserver command tells ypbind to use the named server regardless of what the current NIS domain may be. If, however, you are moving your machine between different NIS domains frequently, you may want to keep information for several domains in the *yp.conf* file. You can have information on the servers for various NIS domains in *yp.conf* by specifying the information using the *domain* statement. For instance, you might change the previous sample file to look like this for a laptop:

```
# yp.conf - YP configuration for GNU libc library.
#
domain winery server vbardolino
domain brewery server vstout
```

This lets you bring up the laptop in either of the two domains by simply setting the desired NIS domain at boot time using the domainname command. The NIS client then uses whichever server is relevant for the current domain.

There is a third option you may want to use. It covers the case when you don't know the name or IP address of the server to use in a particular domain, but still want the ability use a fixed server on certain domains. Imagine we want to insist on using a specified server while operating within the Winery domain, but want to probe for the server to use while in the Brewery domain. We would modify our *yp.conf* file again to look like this instead:

```
# yp.conf - YP configuration for GNU libc library.
#
```

```
domain winery server vbardolino
domain brewery broadcast
```

The *broadcast* keyword tells ypbind to use whichever NIS server it finds for the domain.

After creating this basic configuration file and making sure it is world-readable, you should run your first test to connect to your server. Make sure to choose a map your server distributes, like *hosts.byname*, and try to retrieve it by using the ypcat utility:

```
# ypcat hosts.byname
172.16.2.2   vbeaujolais.vbrew.com    vbeaujolais
172.16.2.3   vbardolino.vbrew.com     vbardolino
172.16.1.1   vlager.vbrew.com         vlager
172.16.2.1   vlager.vbrew.com         vlager
172.16.1.2   vstout.vbrew.com         vstout
172.16.1.3   vale.vbrew.com           vale
172.16.2.4   vchianti.vbrew.com       vchianti
```

The output you get should resemble that just shown. If you get an error message instead that says: Can't bind to server which serves domain, then either the NIS domain name you've set doesn't have a matching server defined in *yp.conf*, or the server is unreachable for some reason. In the latter case, make sure that a ping to the host yields a positive result, and that it is indeed running an NIS server. You can verify the latter by using rpcinfo, which should produce the following output:

```
# rpcinfo -u serverhost ypserv
program 100004 version 1 ready and waiting
program 100004 version 2 ready and waiting
```

8.8 NETWORK FILE SYSTEM

The Network File System (NFS) is probably the most prominent network service using RPC. It allows you to access files on remote hosts in exactly the same way you would access local files. A mixture of kernel support and user-space daemons on the client side, along with an NFS server on the server side, makes this possible. This file access is completely transparent to the client and works across a variety of server and host architectures.

NFS offers a number of useful features:

- Data accessed by all users can be kept on a central host, with clients mounting this directory at boot time. For example, you can keep all user accounts on one host and have all hosts on your network mount */home* from that host. If NFS is installed beside NIS, users can log into any system and still work on one set of files.
- Data consuming large amounts of disk space can be kept on a single host. For example, all files and programs relating to LaTeX and METAFONT can be kept and maintained in one place.
- Administrative data can be kept on a single host. There is no need to use rcp to install the same stupid file on 20 different machines.

It's not too hard to set up basic NFS operation on both the client and server; this chapter tells you how.

Linux NFS is largely the work of Rick Sladkey, who wrote the NFS kernel code and large parts of the NFS serverThe latter is derived from the **unfsd** user space NFS server, originally written by Mark Shand, and the **hnfs** Harris NFS server, written by Donald Becker.

When someone accesses a file over NFS, the kernel places an RPC call to rpc.nfsd (the NFS daemon) on the server machine. This call takes the file handle, the name of the file to be accessed, and the user and group IDs of the user as parameters. These are used in determining access rights to the specified file. In order to prevent unauthorized users from reading or modifying files, user and group IDs must be the same on both hosts.

On most Unix implementations, the NFS functionality of both client and server is implemented as kernel-level daemons that are started from user space at system boot. These are the *NFS Daemon* (rpc.nfsd) on the server host, and the *Block I/O Daemon* (biod) on the client host. To improve throughput, biod performs asynchronous I/O using read-ahead and write-behind; also, several rpc.nfsd daemons are usually run concurrently.

The current NFS implementation of Linux is a little different from the classic NFS in that the server code runs entirely in user space, so running multiple copies simultaneously is more complicated. The current rpc.nfsd implementation offers an experimental feature that allows limited support for multiple servers.

8.9 PREPARING NFS

Before you can use NFS, be it as server or client, you must make sure your kernel has NFS support compiled in. Newer kernels have a simple interface on the *proc* filesystem for this, the */proc/filesystems* file, which you can display using cat:

```
$  cat  /proc/filesystems
            minix
            ext2
            msdos
nodev       proc
nodev       nfs
```

If *nfs* is missing from this list, you have to compile your own kernel with NFS enabled, or perhaps you will need to load the kernel module if your NFS support was compiled as a module.

8.10 MOUNTING AN NFS VOLUME

The mounting of NFS volumes closely resembles regular file systems. Invoke mount using the following syntax:

```
# mount -t nfs nfs_volume local_dir options
```

One doesn't say filesystem because these are not proper filesystems.

nfs_volume is given as *remote_host:remote_dir*. Since, this notation is unique to NFS filesystems, you can leave out the *-t nfs* option.

There are a number of additional options that you can specify to mount upon mounting an NFS volume. These may be given either following the *-o* switch on the command line or in the options field of the */etc/fstab* entry for the volume. In both cases, multiple options are separated by commas and must not contain any whitespace characters. Options specified on the command line always override those given in the *fstab* file.

Here is a sample entry from */etc/fstab*:

```
# volume mount point type options
news:/var/spool/news  /var/spool/news  nfs  timeo=14,intr
```

This volume can then be mounted using this command:

```
# mount news:/var/spool/news
```

In the absence of an *fstab* entry, NFS mount invocations look a lot uglier. For instance, suppose you mount your users' home directories from a machine named *moonshot*, which uses a default block size of 4 K for read/write operations. You might increase the block size to 8 K to obtain better performance by issuing the command:

```
# mount moonshot:/home/home -o rsize=8192,wsize=8192
```

The list of all valid options is described in its entirety in the *nfs(5)* manual page. The following is a partial list of options you would probably want to use:

rsize=n and wsize=n

These specify the datagram size used by the NFS clients on read and write requests, respectively. The default depends on the version of kernel, but is normally 1,024 bytes.

timeo=n

This sets the time (in tenths of a second) the NFS client will wait for a request to complete. The default value is 7 (0.7 seconds). What happens after a timeout depends on whether you use the *hard* or *soft* option.

hard

Explicitly mark this volume as hard-mounted. This is on by default. This option causes the server to report a message to the console when a major timeout occurs and continues trying indefinitely.

soft

Soft-mount (as opposed to hard-mount) the driver. This option causes an I/O error to be reported to the process attempting a file operation when a major timeout occurs.

intr

Allow signals to interrupt an NFS call. Useful for aborting when the server doesn't respond.

Except for *rsize* and *wsize*, all of these options apply to the client's behavior if the server should become temporarily inaccessible. They work together in the following way: Whenever the client sends a request to the NFS server, it expects the operation to have finished after a given interval (specified in the *timeout* option). If no confirmation is received within this time, a so-called *minor timeout* occurs, and the operation is retried with the timeout interval doubled. After reaching a maximum timeout of 60 seconds, a *major timeout* occurs.

By default, a major timeout causes the client to print a message to the console and start all over again, this time with an initial timeout interval twice that of the previous cascade. Potentially, this may go on forever. Volumes that stubbornly retry an operation until the server becomes available again are called *hard-mounted.* The opposite variety, called *soft-mounted,* generate an I/O error for the calling process whenever a major timeout occurs. Because of the write-behind introduced by the buffer cache, this error condition is not propagated to the process itself before it calls the *write* function the next time, so a program can never be sure that a write operation to a soft-mounted volume has succeeded at all.

Whether you hard- or soft-mount a volume depends partly on taste but also on the type of information you want to access from a volume. For example, if you mount your X programs by NFS, you certainly would not want your X session to go berserk just because someone brought the network to a grinding halt by firing up seven copies of Doom at the same time or by pulling the Ethernet plug for a moment. By hard-mounting the directory containing these programs, you make sure that your computer waits until it is able to re-establish contact with your NFS server. On the other hand, non-critical data such as NFS-mounted news partitions or FTP archives may also be soft-mounted, so if the remote machine is temporarily unreachable or down, it doesn't hang your session. If your network connection to the server is flaky or goes through a loaded router, you may either increase the initial timeout using the *timeo* option or hard-mount the volumes. NFS volumes are hard-mounted by default.

Hard mounts present a problem because, by default, the file operations are not interruptible. Thus, if a process attempts, for example, a write to a remote server and that server is unreachable, the user's application hangs and the user can't do anything to abort the operation. If you use the *intr* option in conjuction with a hard mount, any signals received by the process interrupt the NFS call so that users can still abort hanging file accesses and resume work (although without saving the file).

Usually, the rpc.mountd daemon in some way or other keeps track of which directories have been mounted by what hosts. This information can be displayed using the showmount program, which is also included in the NFS server package:

```
# showmount -e moonshot
Export list for localhost:
/home <anon clnt>

# showmount -d moonshot
Directories on localhost:
/home

# showmount -a moonshot
All mount points on localhost:
localhost:/home
```

8.11 THE NFS DAEMONS

If you want to provide NFS service to other hosts, you have to run the rpc.nfsd and rpc.mountd daemons on your machine. As RPC-based programs, they are not managed by inetd, but are started up at boot time and register themselves with the portmapper; therefore, you have to make sure to start them only after rpc.portmap is running. Usually, you'd use something like the following example in one of your network boot scripts:

```
if [ -x /usr/sbin/rpc.mountd ]; then
        /usr/sbin/rpc.mountd; echo -n " mountd"
fi
if [ -x /usr/sbin/rpc.nfsd ]; then
```

```
        /usr/sbin/rpc.nfsd; echo -n " nfsd"
fi
```

The ownership information of the files an NFS daemon provides to its clients usually contains only numerical user and group IDs. If both client and server associate the same user and group names with these numerical IDs, they are said to their share uid/gid space. For example, this is the case when you use NIS to distribute the *passwd* information to all hosts on your LAN.

On some occasions, however, the IDs on different hosts do not match. Rather than updating the uids and gids of the client to match those of the server, you can use the rpc.ugidd mapping daemon to work around the disparity. Using the *map_daemon* option explained a little later, you can tell rpc.nfsd to map the server's uid/gid space to the client's uid/gid space with the aid of the rpc.ugidd on the client. Unfortunately, the rpc.ugidd daemon isn't supplied on all modern Linux distributions, so if you need it and yours doesn't have it, you will need to compile it from source.

rpc.ugidd is an RPC-based server that is started from your network boot scripts, just like rpc.nfsd and rpc.mountd:

```
if [ -x /usr/sbin/rpc.ugidd ]; then
        /usr/sbin/rpc.ugidd; echo -n " ugidd"
```

8.12 THE EXPORTS FILE

Now we'll look at how we configure the NFS server. Specifically, we'll look at how we tell the NFS server what filesystems it should make available for mounting, and the various parameters that control the access clients will have to the filesystem. The server determines the type of access that is allowed to the server's files. The */etc/exports* file lists the filesystems that the server will make available for clients to mount and use.

By default, rpc.mountd disallows all directory mounts, which is a rather sensible attitude. If you wish to permit one or more hosts to NFS-mount a directory, you must *export* it, that is, specify it in the *exports* file. A sample file may look like this:

```
# exports file for vlager
/home vale(rw) vstout(rw) vlight(rw)
/usr/X11R6 vale(ro) vstout(ro) vlight(ro)
/usr/TeX vale(ro) vstout(ro) vlight(ro)
/ vale(rw,no_root_squash)
/home/ftp (ro)
```

Each line defines a directory and the hosts that are allowed to mount it. A hostname is usually a fully qualified domain name but may additionally contain the * and ? wildcards, which act the way they do with the Bourne shell. For instance, lab*.foo.com matches *lab01.foo.com* as well as *laboratory.foo.com*. The host may also be specified using an IP address range in the form *address/netmask*. If no hostname is given, as with the */home/ftp* directory in the previous example, any host matches and is allowed to mount the directory.

When checking a client host against the *exports* file, rpx.mountd looks up the client's hostname using the *gethostbyaddr* call. With DNS, this call returns the client's canonical hostname, so you must make sure not to use aliases in *exports*. In an NIS environment the returned name is the first match from the hosts database, and with neither DNS or

NIS, the returned name is the first hostname found in the *hosts* file that matches the client's address.

The hostname is followed by an optional comma-separated list of flags, enclosed in parentheses. Some of the values these flags may take are:

secure

This flag insists that requests be made from a reserved source port, i.e., one that is less than 1,024. This flag is set by default.

insecure

This flag reverses the effect of the *secure* flag.

ro

This flag causes the NFS mount to be read-only. This flag is enabled by default.

rw

This option mounts file hierarchy read-write.

root_squash

This security feature denies the superusers on the specified hosts any special access rights by mapping requests from uid 0 on the client to the uid 65534 (that is, -2) on the server. This uid should be associated with the user *nobody*.

no_root_squash

Don't map requests from uid 0. This option is on by default, so superusers have superuser access to your system's exported directories.

link_relative

This option converts absolute symbolic links (where the link contents start with a slash) into relative links. This option makes sense only when a host's entire filesystem is mounted; otherwise, some of the links might point to nowhere, or even worse, to files they were never meant to point to. This option is on by default.

link_absolute

This option leaves all symbolic links as they are (the normal behavior for Sun-supplied NFS servers).

map identity

This option tells the server to assume that the client uses the same uids and gids as the server. This option is on by default.

map_daemon

This option tells the NFS server to assume that client and server do not share the same uid/gid space. rpc.nfsd then builds a list that maps IDs between client and server by querying the client's rpc.ugidd daemon.

map_static

This option allows you to specify the name of a file that contains a static map of uids and gids. For example, map_static=/etc/nfs/vlight.map would specify the */etc/nfs/vlight.map* file as a uid/gid map. The syntax of the map file is described in the *exports(5)* manual page.

map_nis

This option causes the NIS server to do the uid and gid mapping.

anonuid and anongid

These options allow you to specify the uid and gid of the anonymous account. This is useful if you have a volume exported for public mounts.

Any error in parsing the *exports* file is reported to syslogd's *daemon* facility at level *notice* whenever rpc.nfsd or rpc.mountd is started up.

8.13 SYSTEM BACKUP & RECOVERY

Performing regular backups should be considered one of a responsible system administrator's top priorities. Although Linux is an extremely reliable operating system, failures can, do, and probably will occur. They may be caused by hardware failure, power outages, or other unforeseen problems.

More likely will be those problems caused by human error, resulting in undesired changes to, or even deletions of, crucial files. If you are hosting users on your system, you will most certainly be requested to restore an inadvertently deleted file or two.

If you perform regular backups, preferably on a daily basis (at least for user files which are updated often), you will hopefully reduce the possibility of, and increase your recovery from, such file loss age.

The safest method of doing backups is to record them on separate media, such as tape, removable drive, writeable CD, etc., and then store your backup sets in a location separate from your Linux system. Sometimes this may not be practical — perhaps you do not have a fire-proof vault in which you can store your backup tapes! Or perhaps you do not have access to such an external backup system in the first place. Nonetheless, backups can still be performed, albeit on a slightly limited basis.

At my place of employment, I perform backups on several Linux servers. Depending on the situation, some of these backup sets are written to tapes; others are written to a separate server over the network, while still others are simply written to a separate disk partition (for example, in the "/archive/" file system) by an automatic cron job (perhaps because the server is in a remote location, for which a daily visit to perform a tape backup is impractical or impossible).

At home, I do not have an external backup system, nor do I have massive amounts of available disk space to write a backup image. Therefore, I instead back up only my user files on "/home/" as well as some customized configuration files in "/etc/", writing the backup set to a separate disk partition.

There are a variety of methods of performing backups with Linux. These include command-line tools included with every Linux distribution, such as "dd", "dump", "cpio", as well as "tar". Also, available are text-based utilities, such as "Amanda" and "Taper", which is designed to add a more user-friendly interface to the backup and restore procedures. There are GUI-based utilities as well, such as "KDat". Finally, commercial backup utilities are also available, such as "BRU" and "PerfectBackup+". Any one of these backup solutions can provide protection for your valuable data.

A brief listing of some of the tools available, including where they can be obtained, can be found on the "Linux Applications and Utilities Page", at when deciding on a backup solution, you will need to consider the following factors:

- *Portability*—Is backup portability (i.e. the ability to backup on one Linux distribution or implementation of Unix and restore to another; for example from Solaris to Red Hat Linux) important to you? If so, you'll probably want to choose one of the command-line tools (e.g. "dd", "dump", "cpio", or "tar"), because you can be reasonably sure that such tools will be available on any *nix system.
- *Unattended or automated backups*—Is the ability to automate backups so that they can be performed at regular intervals without human intervention important to you? If so, you will need to choose both a tool and a backup medium which will support such a backup scheme.
- *User-friendliness*—Is a user-friendly interface important to you? If so, you will likely want to choose a tool which provides a text- or GUI-based interface. The commercial utilities may provide the easiest interfaces as well as added technical support.
- *Remote backups*—Is the ability to start backups and restores from a remote machine important to you? If so, you'll probably want to choose one of the command-line tools or text-based utilities instead of the GUI-based utilities (unless you have a reasonably fast network connection and the ability to run remote X sessions).
- *Network backups*—Is performing backups and restores to and from networked hosts important to you? If so, you'll probably want to use one of several of the command-line utilities (such as "tar") which support network access to backup devices, or a specialized utility such as "Amanda" or one of several commercial utilities.
- *Media types*—Backups can be stored on a variety of medium, such as tape, an extra hard drive, ZIP drives, or rewritable CDs. Consider cost vs. reliability, storage capacitity, and transfer speed.

Caution
Caution: When backing up your file systems, do *not* include the "/proc" pseudo-filesystem! The files in /proc are not actually files but are simply file-like links which describe and point to kernel data structures. Backing up a file like "/proc/kcore", which is actually a pseudo-file containing the contents of your entire memory, seems like a pretty big waste of tape to me! :-) You will also, likely want to avoid backing up the "/mnt" file system, unless you have the peculiar desire to back up the files from your CD-ROM device, floppy drive, network file shares, or other mounted devices.

Obviously, the procedures for performing a backup and restore will differ depending on your choice of a backup solution. However, in this section, I will discuss methods for performing backups with the two tools I use most: "tar" (whose name stands for "Tape ARchiver"), which is a command-line backup tool largely portable across *nix systems; as well as "KDat", a GUI-based tape backup utility which comes included with the KDE packages.

Finally, I should add that, depending on your choice of backup solution, even if the tool doesn't have the ability built-in to schedule automated or unattended backups, you may be able to automate such backups by using the cron facilities.

8.13.1 Backing up with "tar":

If you decide to use "tar" as your backup solution, you should probably take the time to get to know the various command-line options that are available; type "man tar" for a comprehensive list. You will also need to know how to access the appropriate backup media; although all devices are treated like files in the Unix world, if you are writing to a character device such as a tape, the name of the "file" is the device name itself (e.g. "/dev/nst0" for a SCSI-based tape drive).

The following command will perform a backup of your entire Linux system onto the "/archive/" file system, with the exception of the "/proc/" pseudo-filesystem, any mounted file systems in "/mnt/", the "/archive/" file system (no sense backing up our backup sets!), as well as Squid's rather large cache files (which are, in my opinion, a waste of backup media and unnecessary to back up):

```
tar -zcvpf /archive/full-backup-'date '+%d-%B-%Y''.tar.gz \ -directory
/-exclude=mnt  -exclude=proc  -exclude=var/spool/squid.
```

Don't be intimidated by the length of the command above! As we break it down into its components, you will see the beauty of this powerful utility.

The above command specifies the options "*z*" (compress; the backup data will be compressed with "gzip"), "*c*" (create; an archive file is begin created), "*v*" (verbose; display a list of files as they get backed up), "*p*" (preserve permissions; file protection information will be "remembered" so they can be restored). The "*f*" (file) option states that the very next argument will be the name of the archive file (or device) being written. Notice how a filename which contains the current date is derived, simply by enclosing the "date" command between two back-quote characters. A common naming convention is to add a "tar" suffix for non-compressed archives, and a "tar.gz" suffix for compressed ones.

The "*—directory*" option tells tar to first switch to the following directory path (the "/" directory in this example) prior to starting the backup. The "*—exclude*" options tell tar not to bother backing up the specified directories or files. Finally, the "." character tells tar that it should back up everything in the current directory.

NOTE: It is important to realize that the options to tar are cAsE-sEnSiTiVe! In addition, most of the options can be specified as either single mneumonic characters (e.g. "f"), or by their easier-to-memorize full option names (e.g. "file"). The mneumonic representations are identified by prefixing them with a "-" character, while the full names are prefixed with two such characters. Again, see the "man" pages for information on using tar.

Another example, this time writing only the specified file systems (as opposed to writing them *all* with exceptions as demonstrated in the example above) onto a SCSI tape drive follows:

```
tar -cvpf /dev/nst0 -label="Backup set created on 'date '+%d-%B-
%Y''." \ -directory / -exclude=var/spool/ etc home usr/local var/
spool
```

In the above command, notice that the "*z*" (compress) option is not used. I *strongly* recommend against writing compressed data to tape, because if data on a portion of the tape becomes corrupted, you will lose your entire backup set! However, archive files stored without compression have a very high recoverability for non-affected files, even if portions of the tape archive are corrupted.

Because the tape drive is a character device, it is not possible to specify an actual file name. Therefore, the file name used as an argument to tar is simply the name of the device, "/dev/nst0", the first tape device on the SCSI bus.

NOTE: The "/dev/nst0" device does not rewind after the backup set is written; therefore it is possible to write multiple sets on one tape. (You may also refer to the device as "/dev/st0", in which case the tape is automatically rewound after the backup set is written.)

Since, we aren't able to specify a filename for the backup set, the "*—label*" option can be used to write some information about the backup set into the archive file itself.

Finally, only the files contained in the "/etc/", "/home/", "/usr/local", and "/var/spool/" (with the exception of squid's cache data files) are written to the tape.

When working with tapes, you can use the following commands to rewind, and then eject your tape:

```
mt -f /dev/nst0 rewind
mt -f /dev/nst0 offline
```

Tip: You will notice that leading "/" (slash) characters are stripped by tar when an archive file is created. This is tar's default mode of operation, and it is intended to protect you from overwriting critical files with older versions of those files, should you mistakenly recover the wrong file(s) in a restore operation. If you really dislike this behavior (remember, its a *feature* !) you can specify the "—absolute-paths" option to tar, which will preserve the leading slashes. However, I don't recommend doing so, as it is *Dangerous* !.

8.13.2 Backing up with "KDat":

If you are using the KDE desktop environment, I believe you will find the "KDat" utility both powerful as well as user-friendly. In addition, an added bonus is that KDat uses "tar" as its backup engine. Therefore, backup sets written with KDat can be read not only with KDat but with tar as well! This makes KDat a very nice choice for both user-friendliness as well as backup portability.

Tip: Even if you choose not to use nor install the full KDE package, you can *still* use KDat as long as you have the Qt libraries installed.

The first time you run the KDat program, you will need to create a backup profile. Such a profile tells KDat which files on your system you would like to back up. If you wish, you can create more than one backup profile, depending on your needs (for example, you could create a profile called "*Full Backup*" for a full system backup, and "*Quick Backup*" for a backup of user files only).

To create a backup profile, either choose "Create Backup Profile" from the "File" option on menu bar (or right-click on the "Backup Profiles" folder, then choose "Create Backup Profile"). On the right hand side of the KDat window, you can change various settings, such as the profile name, archive name, tar options, as well as others. Click the "Help" menu for more information on what these settings are for.

To specify which files should be included in your backup profile, left-click the checkbox beside the "/" directory folder. This will enable all files in and below this directory for backups. Then, left-click on the small "+" sign beside the folder. This will expand the folder, showing a list of files in and below it. This will allow you to exclude any files you do not wish to backup; simply left-click the checkbox beside each file or directory you wish to exclude. For example, a full backup should probably have every file and directory checkmarked, with the exception of the "/proc" (a pseudo-filesystem containing information about your running system), "/mnt" (a directory below which CD-ROM drives, floppies, and network shares are usually mounted), and, if you are a Squid user, "/var/spool/

squid" (Squid's cache data files). Once you have selected the appropriate files, left-click on the backup profile you are creating, then left-click the "Files >>" button to move the selected files list to your backup profile.

NOTE : Should your server data be larger in size than can be physically stored on a tape, you will need to create separate backup profiles, one for each portion of your backup set.

To actually perform a backup, insert a tape into the drive, and then choose "Mount Tape" from the "File" menu (or left-click the icon that looks like a tape). This will "*mount*" the tape (actually, because a tape device is a character device, it isn't actually possible to mount it — what KDat actually does is to first rewind the tape, attempts to read in header information, and if successful, find the corresponding tape index on your hard drive. Otherwise, KDat will prompt you to format the tape.

NOTE : (Note: If KDat keeps complaining that a tape isn't in the drive and it actually *is* in the drive, you should ensure the correct tape device name is specified in the preferences; left-click the "Edit" option on the menu bar and choose "User Preferences".)

Once KDat has mounted the tape, before you start the backup you must first choose the backup profile you wish to use for the backup. To start the backup, simply, right-click on the desired backup profile, and then left-click on the "Backup" option. KDat will first display a dialog box showing you the details of the backup profile you have selected; left-click the "Ok" button to start the backup.

While the backup is in progress, KDat will display a dialog box showing various statistical information (elapsed time, backup size, backup rate, estimated time remaining, as well as the number of files and total bytes written), and display a list of files as they are backed up. A full backup containing several gigabytes of data might take several hours to complete. If you find it necessary, you can left-click the "Abort" button at any time to interrupt the backup process.

Once the backup is complete, you can unmount the tape by choosing "Edit" from the menu bar, and then "Unmount Tape", or left-click on the tape icon, which will rewind and eject the tape.

8.14 LOG FILES FOR SYSTEM AND APPLICATIONS

When compiling the UUCP suite to use Taylor-style logging, you have only three global log files, all of which reside in the spool directory. The main log file is named *Log* and contains all the information about established connections and transferred files. A typical excerpt looks like this (after a little reformatting to make it fit the page):

```
uucico pablo - (1994-05-28 17:15:01.66 539) Calling system pablo (port
cua3)
uucico pablo - (1994-05-28  17:15:39.25  539)  Login  successful
uucico pablo - (1994-05-28  17:15:39.90  539)  Handshake  successful
               (protocol 'g' packet size 1024 window 7)
uucico pablo postmaster (1994-05-28 17:15:43.65 539) Receiving D.pabloB04aj
uucico pablo postmaster (1994-05-28 17:15:46.51 539) Receiving X.pabloX04ai
uucico pablo postmaster (1994-05-28 17:15:48.91 539) Receiving D.pabloB04at
uucico pablo postmaster (1994-05-28 17:15:51.52 539) Receiving X.pabloX04as
```

```
uucico pablo postmaster (1994-05-28 17:15:54.01 539) Receiving D.pabloB04c2
uucico pablo postmaster (1994-05-28 17:15:57.17 539) Receiving X.pabloX04c1
uucico pablo - (1994-05-28 17:15:59.05 539) Protocol 'g' packets: sent 15,
            resent 0, received 32
uucico pablo - (1994-05-28 17:16:02.50 539) Call complete (26 seconds)
uuxqt pablo postmaster (1994-05-28 17:16:11.41 546) Executing X.pabloX04ai
            (rmail okir)
uuxqt pablo postmaster (1994-05-28 17:16:13.30 546) Executing X.pabloX04as
            (rmail okir)
uuxqt pablo postmaster (1994-05-28 17:16:13.51 546) Executing X.pabloX04c1
            (rmail okir)
```

The next important log file is *Stats*, which lists file transfer statistics. The section of *Stats* corresponding to the above transfer looks like this (again, the lines have been split to fit the page):

```
postmaster pablo (1994-05-28 17:15:44.78)
            received 1714 bytes in 1.802 seconds (951 bytes/)
postmaster pablo (1994-05-28 17:15:46.66)
            received 57 bytes in 0.634 seconds (89 bytes/)
postmaster pablo (1994-05-28 17:15:49.91)
            received 1898 bytes in 1.599 seconds (1186 bytes/)
postmaster pablo (1994-05-28 17:15:51.67)
            received 65 bytes in 0.555 seconds (117 bytes/)
postmaster pablo (1994-05-28 17:15:55.71)
            received 3217 bytes in 2.254 seconds (1427 bytes/)
postmaster pablo (1994-05-28 17:15:57.31)
            received 65 bytes in 0.590 seconds (110 bytes/)
```

The third file is *Debug*. Debugging information is written here. If you use debugging, make sure this file has protection mode 600. Depending on the debug mode you select, it may contain the login and password you use to connect to the remote system.

If you have some tools around that expect your log files to be in the traditional format used by HDB-compatible UUCP implementations, you can also, compile Taylor UUCP to produce HDB-style logs. This is simply a matter of enabling a compile-time option in *config.h*

Viewing System Log Entries

The following two procedures describe how to access and view system logs.

To view system log entries

1. From the VI Client connected to either a VirtualCenter Server or an ESX Server. display the administration panel and view the system log tab.
2. Click the Admin button in the navigation bar. Click the System Logs tab.
3. From the drop-down list, choose the log and entry you want to view.

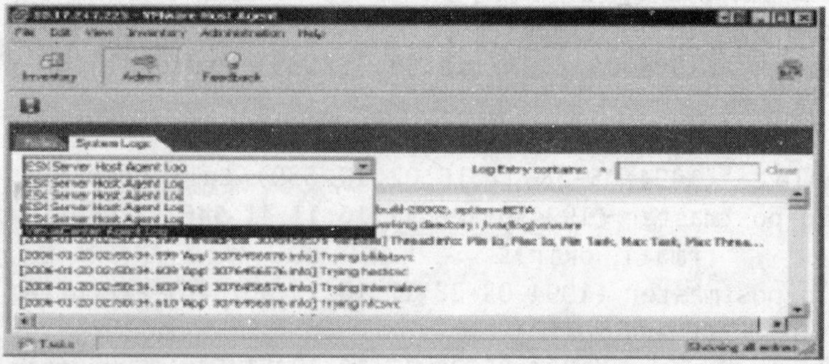

To search and list selected items

1. With the log you want to filter displayed, choose View>Filtering to refer to the filtering options.
2. Enter text into the data field. Click Clear to empty the data field.

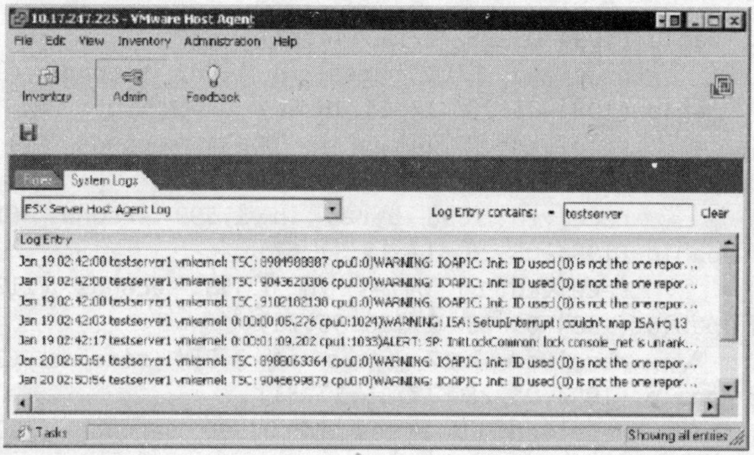

External System Logs

VMware technical support might request several files to help resolve any issues you have with the product. The following table describes additional log files you might need.

Component	Operating System	File Name	Location
ESX Server 2.x Service log	ESX Server	vmware-serverd.log.	/var/logvmware/
ESX Server 3.x Service log	ESX Server	hostd.log	/var/log/vmware/
VI Client Agent log	ESX Server	vpxa.log	/var/log/vmware/vpx/
VI Client Installation log	Windows		Virtual Infrastructure Client host.
VI Client Service log	Windows	temp_dir/viclient-x-log x(=0,1, ...9)	Temp directory

Component	Operating System	File	Name Location
Virtual Machine Console logs	Windows	vmware-<username> <PID>, log	Temp directory Example: C:\Documents and Settings\<username>\Local Setting\Temp The path to this file appears in the About dialog box. In a console, choose Help > About VMware Virtual Infrastructure Client, and look under Additional Information.
	Linux	ui-<PID>.log	Temp directory Example: /tmp/vnware-<username>/ The path to this file appears in the terminal when you start the console.
	If you encounter problems with the VMware Virtual Machine console on a remote VI Client, please submit a support request and this log file.		
Virtual Machine Console Installation log	Windows	VMInst.log	Temp directory Example: C:\Documents and Settings\<username>\Local Settings\Temp. The Local Settings folder is hidden by default. To refer to its contents, open My Computer, choose Tools > Folder Options, click the View tab, and choose Show Hidden Files and Folders.
	Linux	VMInst.log	/etc/vnware/
Virtual Machine log	Windows	vmware.log	In the same directory as the configuration file (.vmx) of the virtual machine that had problems. The path to the log file of the active virtual machine appears in the About dialog box. In a console, choose Help > About VMware Virtual Infrastructure Client, and look under Additional information.
	Linux	vmware.log	In the same directory as the configuration file < .vmx) of the virtual machine that had problems.
	Run the support script or save the log file before you launch the failed virtual machine again.		

Component	Operating System	File	Name Location
Virtual Machine's Even log	Window	event-<path_to_ configuretion_file> .vmx.log	On die virtual machine managed host. C:\Program Files\VMware\ VMware Virtual Infrastructure\ vmserverdRoot\cventlog
	Linux		/var/log/vmware
Virtual Machine Kernel Core file	ESX Server	vmkernel-core.<date> vrakerngl-log.<date>	/root directory after you reboot your machine.
Virtual Machine Configuration file	Windows and Linux	.vmx	On the virtual machine managed host.
Service Console log	ESX Server		/var/log/messages
Service Console Availability report	ESX Server		
VMkemel Messages	ESX Server		/var/log/vmkernel in the service console

8.15 BACKUP SCHEDULES AND METHODS (MANUAL AND AUTOMATED)

To back up your data, you can use one of the archiving tools that are already available on your system. Tar, cpio, dump are only some of these tools. But take into account that you have a lot to do to use one of these tools comfortably.

You should consider using one of the third-party products we mentioned earlier (BRU, Arkeia, or BackupEDGE/RecoverEDGE) for your backup tool. They come with a command line interface, so you have full control over the tools' functionality, and you also have the use of a graphical interface. These products should include a scheduler to make it easier for you to set up the periodic scheduling scheme most comfortable for you.

Information on the commercially available backup solutions can be found at the following Web sites:

```
Arkeia:                    http://www.arkeia.con
BRU/CRU:                   http://www.estinc.cem
BackupEDGE/RecoverEDGE:    http://wtM.microiite.can
```

8.15.1 BRU (Backup and Recovery Utility)

It is the backup and restore solution from

Enhanced Software Technologies. It is a backup and restore utility with significant enhancements over other common utilities such as tar, cpio, volcopy and dump. BRU is designed to work with most backup devices, including cartridge, 4 mm DAT, 8 mm (Exabyte) and 9-track tape drives.

BRU includes incremental backups, full backups, multivolume archives, distribution and updates, error detection and recovery, random access capabilities, file comparisons, and file overwrite protection.

As part of the package, BRU also contains a graphical user interface (GUI) named XBRU. Used from the command line, BRU is very similar to Tar. The dialog windows of XBRU are very intuitive and easy to use. To set up a scheme for scheduled backups, it is easiest to use the BRU for X11 scheduler component of XBRU.

CRU (Crash Recovery Utility) is the recovery solution from Enhanced Software Technologies. It allows you to recover your system after a disaster crash.

To use CRU, you must create two floppy disks (a boot disk and a root disk) and/or a bootable CD-ROM, which are used to start the system recovery after a disaster crash. Follow the instructions in the CRU documentation to create the boot media. After every change in the system layout, mainly the layout of the hard disks, you have to repeat the creation of the CRU boot and root disks.

CRU requires that you create a special full backup tape using CRU. This backup feature of CRU can replace the corresponding BRU feature to create full backups.

8.15.2 BackupEDGE and RecoverEDGE

BackupEDGE is the backup and restore solution from MicroLite. It supports SCSI and ATAPI tapes as well as some tape libraries. For more information, check MicroLite's Web page and search on

Device Compatibility to obtain a list of the supported tape drives and autochangers.

RecoverEdge is the recovery solution from MicroLite. It allows you to recover your system after irreparable damage has been done to the system disk.

You have to create some floppy disks (boot disk, root disk, and maybe a second root disk) and/or a bootable CD-R, which are used to boot from to begin a disaster recovery of your system. It is necessary to create these disks after every change in your system configuration, mainly after every change in the layout of the system disks. Following a well-planned backup scheme is also a prerequisite.

To recover after a disaster crash, you need the last full backup media and all subsequent incremental backup media created with BackupEDGE.

8.15.3 Arkeia

Arkeia is a backup and restore solution that allows centralized network backup in a client/server architecture. With Arkeia, you can also back up UNIX, Windows, Windows NT or Novell clients.

With Arkeia you can safely archive every file, directory, device node and special file on your file systems, unlike the standard UNIX tar command, which ignores many important files. Arkeia also verifies the data written to tape to ensure that the tape is an accurate reflection of your data. The following are features provided by Arkeia backup software:

- Data compression - automatic data compression is supported.
- GUI and CLI interface.
- The backup server may be your local system or a remote system.
- High performance - advanced double buffering and variable block factors.

- Virtual file support - you can back up virtual (sparse) files.
- Multi-volume/multi-device archives - automatic spanning across multiple volumes or devices.
- Wildcard support - when selecting files you can use a wildcard.
- Raw device backups - you can archive an entire raw device/partition to tape.
- Master/incremental backups.
- Unattended operation - you can configure schemes to periodically perform full backups and/or incremental backups.

Arkeia is designed to operate on Linux kernels 2.x and there are versions available for several types of libraries (libc5 and libc6) and distributions.

Arkeia supports only devices connected to SCSI-controllers and also tape libraries.

SUMMARY

1. This is why Sun developed the *Network Information System* (NIS). NIS provides generic database access facilities that can be used to distribute, for example, information contained in the *passwd* and *groups* files to all hosts on your network.
2. This makes the network appear as a single system, with the same accounts on all hosts. Similarly, you can use NIS to distribute the hostname information from */etc/hosts* to all machines on the network.
3. NIS is based on RPC, and comprises a server, a client-side library, and several administrative tools. Originally, NIS was called *Yellow Pages*, or YP, which is still used to refer to it.
4. Unfortunately, the name is a trademark of British Telecom, which required Sun to drop that name. As things go, some names stick with people, and so YP lives on as a prefix to the names of most NIS-related commands such as ypserv and ypbind.
5. Today, NIS is available for virtually all Unixes, and there are even free implementations. BSD Net-2 released one that has been derived from a public domain reference implementation donated by Sun.
6. When you're running a local area network, your overall goal is usually to provide an environment for your users that makes the network transparent.
7. An important stepping stone is keeping vital data such as user account information synchronized among all hosts. This provides users with the freedom to move from machine to machine without the inconvenience of having to remember different passwords and copy data from one machine to another.
8. Data that is centrally stored doesn't need to be replicated, so long as there is some convenient means of accessing it from a network-connected host.
9. We previously discussed an important example of this concept that is used on the Internet — the Domain Name System (DNS). DNS serves a limited range of information, the most important being the mapping between hostname and IP address.
10. For other types of information, there is no such specialized service. Moreover, if you manage only a small LAN with no Internet connectivity, setting up DNS may not seem to be worth the trouble.

SHORT QUESTIONS WITH ANSWER

1. Display the /var/run/utmp file.

Ans. who

2. Display the /var/log/wtmp file.

Ans. last

3. Use the lastlog and lastb commands, understand the difference.

Ans.
```
lastlog : when users last logged on
lastb: failed (bad) login attempts
```

4. Examine syslog to find the location of the log file containing ssh failed logins.

Ans.
```
rooterhel53 ~# grep authpriv /etc/syslog.conf
authpriv.*      /var/log/secure
Debian/Ubuntu: /var/log/auth.log
```

Ubuiitii 9.10 and Debian Lenny have switched to usine **rsyslog.**

```
root@ubuntu910 : -# grep authpriv /etc/rsyslog.d/50-default.conf
auth,authpriv.* /var/log/auth.log
root@deb503: -# grep authpriv /etc/rsyslog.conf
auth, authpriv.* /var/log/auth.log
```

5. Configure syslog to put Iocal4.error and above messages in /var/log/14e.log and loca4.infoonly .info in var/log. 14i.log. Test that it works with the logger tool!

Ans.
```
echo Iocal4.error /var/log/14e.log >> /etc/syslog.conf
echo local4.=info /var/log/14i.log >> /etc/syslog.conf
/etc/init.d/syslog restart
logger -p local4.error "14 error test"
logger -p local4.alert "14 alert test"
logger -p local4.info "14 info test"
cat. /var/log/l4e.log
cat /var/log/14i.log
```

6. Configure /var/log/Mysu.log. all the su to root messages should go in that log. Test that it works!

Ans.
```
echo auchpriv.* /var/log/Mysu.log >> /etc/syslog.conf
```

This will log more than just the **su** usage.

7. Send the locals messages to the syslog server of your neighbour. Test that it work On RHEL5, edit /etc/sysconfig/syslog to enable remote listening on the server. On Debian/Ubuntu edit /etc/default/syslog or /etc/default/rsyslog.

Ans.
```
on the client: logger -p locals.into "test locals to neighbour"
```

8. Write a script that executes logger to loca4 every 15 seconds (different message Use tail -f and watch on your Iocal4 log files.

Ans.
```
root@rhel53 scripts# cat logloop
#!/bin/bash
for i in 'scq 1 10"
do
loyger -p Iocal4.info "local4.into test number $i"
sleep 15
done
root@rhel53 scripts# chraod +x logloop
root@rhel53 scripts# ./logloop &
[1] 8264
root@rhel53 scripts# tail -f /var/log/local4.all.log
Mar 28 13:13:36 rhel53 root: Iocal4.info test number 1
Mar 28 13:13:51 rhel53 root: Iocal4.info test number 2
```

9. Use dmesg to find the total amount of memory in your computer.

Ans.
```
dmesg | grep Memory
```

10. Use free to display memory usage in kilobytes (then in megabytes).

Ans. `ftree ; free -m`

11. On a virtual machine, create a swap partition (you might need an extra virtual disk for this).

Ans. `mkswap /dev/sdd1 ; swapon /dev/sdd1`

12. Add a 20 megabyte swap file to the system.

Ans.
```
dd if=/dev/zero of=/swapfile20mb bs=1024 count=20000
mkswap /swapfile20mb
swapon /swapfile20mb
```

13. Put all swap spaces in /etc/fstab and activate them. Test with a reboot that they are mounted.

Ans.
```
root@computert# tail -2 /etc/fstab
/dev/sdd1 swap swap defaults 0 0
/swapfile20mb swap swap defaults 0 0
```

14. Use free to verify usage of current swap.

Ans. `free -om`

EXERCISE

1. Create a volume group that contains a complete disk and a partition on another disk.
2. Create two logical volumes (a small one and a bigger one) in this volumegroup. Format them wih ext3, mount them and copy some files to them.
3. Verify usage with fdisk, mount, pvs, vgs, lvs, pvdisplay, vgdisplay, lvdisplay and df. Does fdisk give you any information about lvm?
4. Enlarge the small logical volume by 50 percent, and verify your work!
5. Take a look at other commands that start with vg* , pv* or lv*.
6. Create a mirror and a striped Logical Volume.
7. Convert a linear logical volume to a mirror.
8. Convert a mirror logical volume to a linear.
9. Create a snapshot of a Logical Volume, take a backup of the snapshot. Then delete some files on the Logical Volume, then restore your backup.
10. Move your volume group to another disk (keep the Logical Volumes mounted).
11. If time permits, split a Volume Group with vgsplit, then merge it again with vgmerge.

❒❒❒

CHAPTER 9

ACTIVE DIRECTORY AND LDAP

9.0 INTRODUCTION

Chapter Outline

There are many areas in which Windows 2000 security is adapting to support the Internet-based enterprise. Some of these changes reflect advances in supporting large organizations through the use of the hierarchical Windows 2000 Active Directory. Other changes take advantage of the flexibility of the Windows security architecture to integrate authentication using Internet public-key certificates.

The list below introduces the new Windows 2000 security features:

- The Active Directory provides the store for all domain security policy and account information. The Active Directory, which provides replication and availability of account information to multiple Domain Controllers, is available for remote administration.
- The Active Directory supports a hierarchical name space for user, group, and computer account information. Accounts can be grouped by organizational units, rather than the flat domain account name space provided by earlier versions of Windows NT.
- Administrator rights to create and manage user or group accounts can be delegated to the level of organizational units. Access rights can be granted to individual properties on user objects to allow, for example, a specific individual or group to have the right to reset passwords, but not to modify other account information.
- Active Directory replication allows account updates at any domain controller, not just the primary domain controller (PDC). Multiple master replicas of the Active Directory at other

domain controllers, which used to be known as backup domain controllers (BDCs), are updated and synchronized automatically.

- Windows 2000 employs a new domain model that uses the Active Directory to support a multilevel hierarchy tree of domains. Management of trust relationships between domains is simplified through tree-wide transitive trust throughout the domain tree.
- Windows security includes new authentication based on Internet standard security protocols, including Kerberos Version 5 and Transport Layer Security (TLS) for distributed security protocols, in addition to supporting Windows NT LAN Manager authentication protocols for compatibility.
- The implementation of secure channel security protocols (SSL 3.0/TLS) supports strong client authentication by mapping user credentials in the form of public-key certificates to existing Windows NT accounts. Common administration tools are used to manage account information and access control, whether using shared secret authentication or public-key security.
- Windows 2000 supports the optional use of smart cards for interactive logon, in addition to passwords. Smart cards support cryptography and secure storage for private keys and certificates, enabling strong authentication from the desktop to the domain.

Windows 2000 provides the Microsoft Certificate Server for Organizations to issue X.509 version 3 certificates to their employees or business partners. This includes the introduction of the CryptoAPI for certificate management and modules to handle public-key certificates, including standard format certificates issued by either a commercial Certificate Authority (CA), third-party CA, or the Microsoft Certificate Server included in Windows. System administrators define which CAs are trusted in their environment and, therefore, which are :

- Certificates are accepted for client authentication and access to resources.
- External users who do not have Windows 2000 accounts can be authenticated using public-key certificates and mapped to an existing Windows account. Access rights defined for the Windows account determine the resources that the external users can use on the system. Client authentication, using public-key certificates, allows Windows 2000 to authenticate external users, based on certificates issued by trusted Certificate Authorities.
- Windows 2000 users have easy-to-use tools and common interface dialogs for managing the private/public-key pairs and the certificates that they use to access Internet-based resources. Storage of personal security credentials, which uses secure disk-based storage, is easily transported with the proposed industry-standard protocol, Personal Information Exchange. The operating system also has integrated support for smart card devices.
- Encryption technology is engineered into the operating system in many ways to take advantage of the use of digital signatures for providing authenticated data streams. In addition to signed ActiveX™ controls and Java Classes for Internet Explorer, Windows 2000 uses digital signatures for image integrity of a variety of program components. In-house developers can also create signed software for distribution and virus protection.

In addition to these changes, we expect third parties to host dynamic password authentication services on Windows 2000 Server and to integrate dynamic passwords with Windows 2000 domain authentication. The APIs and documentation to support these third-party products are available in the Microsoft Platform SDK.

Each of the new features of Windows 2000 security is described in more detail in the following sections.

9.1 ACTIVE DIRECTORY AND SECURITY

Windows NT account information is maintained today using a secure portion of the registry on the domain controllers. Using domain trust and pass-through authentication, a two-level hierarchy of domains provides some flexibility for organizing account management and resource servers. Within a domain, however, accounts are maintained in a flat-name space with no internal organization.

Windows 2000 Distributed Security Services use the Active Directory as the repository for account information. The Active Directory provides significant improvement over the registry-based implementation in the areas of performance and scalability, and offers a feature-rich administrative environment.

The following diagram shows the hierarchical structure for a tree of Windows 2000 domains, and the hierarchical name context within each domain using organizational units (OUs) as directory object containers.

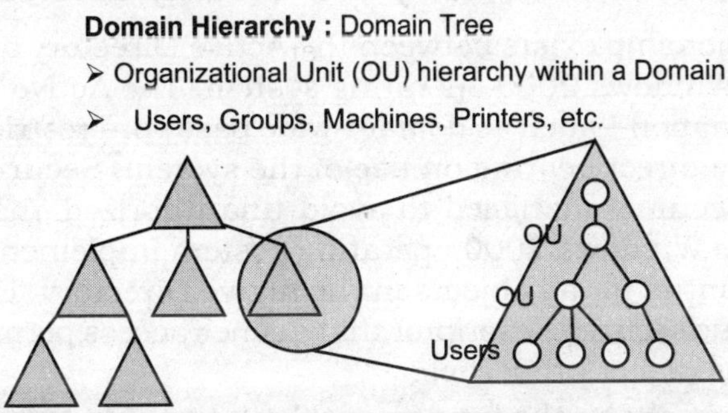

Fig. 9.1. Hierarchical structure of the Active Directory

9.1.1 Advantages of Active Directory Account Management

The advantages of integrating security account management with the Active Directory are:

- Accounts for users, groups, and machines can be organized into directory containers called organizational units (OUs). A domain can have any number of OUs organized in a tree-structured name space. Businesses can organize the name space for account information to represent the departments and organizations in the company. User accounts, as well as OUs, are directory objects that can easily be renamed within the domain tree as the organization changes.
- The Active Directory supports a much larger number of user objects (more than 1 million objects) with better performance than the registry. Individual domain size is no longer limited by performance of the security account repository. A tree of connected domains can support much larger, complex organizational structures.
- Administration of account information is enhanced using advanced graphical tools for Active Directory management, as well as through OLE DS support for scripting languages. Common tasks can be implemented using batch scripts to automate administration.
- Directory replication services support multiple copies of account information in which updates can be made at any copy, not just the designated primary domain controller.

The Lightweight Directory Access Protocol (LDAP) and directory synchronization support provide the mechanisms to link the Windows directory with other directories in the enterprise.

Storing the security account information in the Active Directory means that users and groups are represented as objects in the directory. Read and write access to objects in the directory can be granted to the object as a whole, or to individual properties of the object. Administrators have fine-grain control over who can update user or group information. For example, a telecom operator group can be granted write access only to user account properties related to office telephone equipment without requiring full Account Operator or Administrator privileges.

The concept of a group is also, simplified because local and global groups are both represented by *group objects* in the directory. Existing programming interfaces for local group access are still supported for complete backward compatibility. However, groups defined in the directory can be used for domain-wide access control to resources or only for local administration purposes on the domain controller.

9.1.2 Relationship between Directory and Security Services

A fundamental relationship exists between the Active Directory and the Security Services integrated into the Windows 2000 operating system. The Active Directory stores domain security policy information—such as domain-wide password restrictions and system access privileges—that have direct bearing on use of the system. Security-related objects in the directory must be securely managed to avoid unauthorized changes that affect overall system security. The Windows 2000 operating system implements object-based security model and access control for all objects in the Active Directory. Every object in the Active Directory has a unique security descriptor that defines access permissions that are required to read or update the object properties.

The diagram below shows the fundamental relationship between the Active Directory and operating system security services.

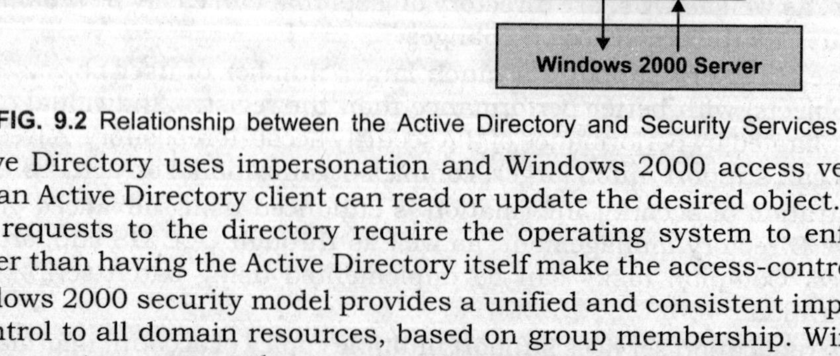

FIG. 9.2 Relationship between the Active Directory and Security Services

The Active Directory uses impersonation and Windows 2000 access verification to determine if an Active Directory client can read or update the desired object. This means LDAP client requests to the directory require the operating system to enforce access control, rather than having the Active Directory itself make the access-control decisions.

The Windows 2000 security model provides a unified and consistent implementation of access control to all domain resources, based on group membership. Windows 2000 security components can trust the security related information stored in the directory. For example, the Windows 2000 authentication service stores encrypted password information in a secure portion of the directory user objects. The operating system trusts that security policy information is stored securely and that account restrictions or group membership is not changed by anyone without authorized access. In addition, security policy information for overall domain management is kept in the directory.

This fundamental relationship of Security and the Active Directory is achieved only by complete integration of the directory with the Windows 2000 operating system and is not otherwise available.

9.1.3 Domain Trust Relationships

Windows 2000 domains can be organized into a hierarchical domain tree. The trust relationships between domains allow users with accounts defined in one domain to be authenticated by resource servers in another domain. In Windows NT 4.0 and earlier versions, interdomain trust relationships are defined by one-way trusted domain accounts between domain controllers. Management of the trust relationships between account domains and resource domains on a large network is a complex task.

The Active Directory supports two forms of trust relationships:

- Explicit one-way trust relationships to Windows NT 4.0 domains.
- Two-way transitive trust between domains that are part of the Windows 2000 domain tree.

The diagram below shows the two styles of trust relationship.

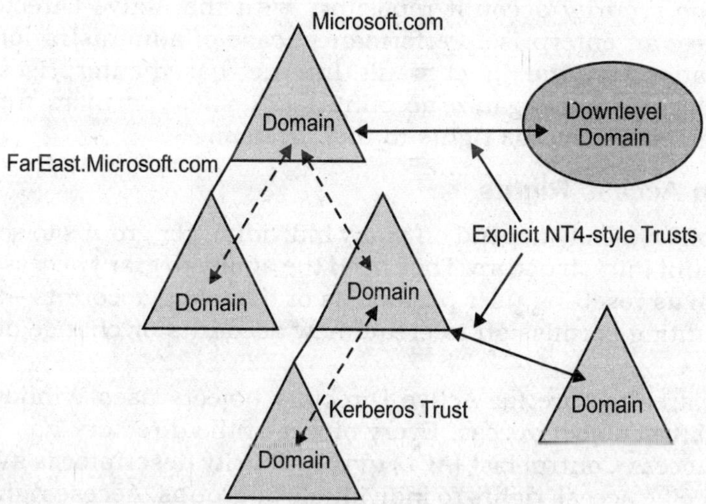

FIG. 9.3 Domain trust relationships

Transitive trust between domains simplifies the management of interdomain trust accounts. Domains that are members of the domain tree define a two-way trust relationship with the parent domain in the tree. All domains implicitly trust other domains in the tree. If there are specific domains that do not want two-way trust, explicit one-way trust accounts can be defined. For organizations with multiple domains, the overall number of explicit one-way trust relationships is significantly reduced.

9.1.4 Delegation of Administration

Delegation of administration is a valuable tool for organizations to confine the security administration to apply only to defined subsets of the entire organization domain. The important requirement is to grant rights to manage a small set of users or groups within their area of responsibility and, at the same time, not give permissions to manage accounts in other parts of the organization.

Delegation of responsibility to create new users or groups is defined at the level of an organizational unit (OU), or container, where the accounts are created. Group administrators for one organizational unit do not necessarily have the ability to create and manage accounts for another organizational unit within a domain. However, domain-wide policy settings and access rights defined at higher levels in the directory tree can apply throughout the tree using inheritance of access rights.

There are three ways to define the delegation of administration responsibilities:

- Delegate permissions to change properties on a particular container, such as the LocalDomainPolicies of the domain object itself.
- Delegate permissions to create and delete child objects of a specific type beneath an OU, such as Users, Groups, or Printers.
- Delegate permissions to update specific properties on child objects of a specific type beneath an OU; for example, the right to set passwords on User objects.

The Directory Service Administration user interface makes it easy to view the delegation information defined for containers. Adding new delegation of permissions is also easy to do by selecting whom you want to delegate permission to, and then choosing which permissions they need.

Integrating the security account repository with the Active Directory provides real benefits to manage an enterprise. Performance, ease of administration, and scalability for large organizations are the direct result. Internet-based enterprises can use domain trees and hierarchical OUs to organize accounts for business partners, frequent customers, or suppliers with specific access rights to their system.

9.1.5 Fine-Grain Access Rights

Large organizations typically depend on many individuals or groups to secure and manage the network account infrastructure. They need the ability to grant access rights for specific operations—such as resetting user passwords or disabling accounts—to specific groups without also, granting permission to create new accounts or change other properties of user accounts.

The security architecture for Active Directory objects uses Windows 2000 security descriptors to control object access. Every object in the directory has a unique security descriptor. The Access Control List (ACL) in the security descriptor is a list of entries that grant or deny specific access rights to individuals or groups. Access rights can be granted or denied with different levels of scope on the object. Access rights can be defined on any of the following levels:

- Apply to the object as a whole, which applies to all properties of the object.
- Apply to a grouping of properties defined by property sets within the object.
- Apply to an individual property of the object.

Granting uniform read/write access to all properties of an object is the default access permission for the creator of the object. Granting or denying object access permissions to a property set is a convenient way to define permissions for a group of related properties. The grouping of properties is defined by the property set attribute of a property in the schema. Changing the schema can customize the property set relationship. Finally, the definition of access rights on a per-property level provides the highest level of granularity of permissions. Definition of per-property access is available on all objects in the Active Directory.

Container objects in the directory also, support fine-grain access with respect to who has permissions to create child objects and what type of child objects they may create. For example, the access control defined on an organizational unit (OU) can define who is allowed to create User objects (accounts) in this container. Another entry in the access control for the OU might define who is allowed to create Printer objects. Fine-grain access control on directory containers is an effective way to maintain organization of the directory name space.

A new implementation of the Access Control List (ACL) Editor, the common dialog control for viewing or changing object security permissions, provides an easy-to-use interface for defining access rights to Active Directory objects by property set or individual properties. The ACL Editor also supports defining inherited access rights on container objects that flow down to all subobjects in that portion of the directory tree.

9.1.6 Inheritance of Access Rights

Inheritance of access rights refers to how access control information defined at higher level containers of the directory flows down to subcontainers and leaf objects. There are generally two models for implementing inherited access rights: dynamic and static inheritance. Dynamic inheritance determines the effective access rights to an object by evaluating the permissions defined explicitly on the object and those defined for all parent objects in the directory. This allows flexibility to change access control on portions of the directory tree by making changes to a specific container that automatically affects all subcontainers and leaf objects. The trade-off to this flexibility is the performance cost to evaluate effective access rights at the time a client requests a read/write to a specific directory object.

Windows 2000 implements a static form of inheritance of access rights, referred to as *Create Time* inheritance. Access control information that flows down to child objects of the container can be defined. When the child object is created, the inherited rights from the container are merged with default access rights on the new object. Any changes to inherited access rights at higher levels in the tree must be propagated down to all affected child objects. New inherited access rights are propagated by the Active Directory to objects for which they apply, based on options for how the new rights are defined.

Performance for access control verification is very fast, using the static model of inheritance of access rights. Access checks are frequent and necessary operations that the operating system is designed to optimize—not just for directory object access, but also for the file system and all other Windows 2000 system objects.

9.2 MULTIPLE SECURITY PROTOCOLS

Windows 2000 supports multiple network security protocols because each protocol provides either compatibility for existing clients, more effective security mechanisms, or interoperability features for heterogeneous networks like the Internet. There are many authentication protocols in use in corporate networks today, and the Windows 2000 architecture does not limit which protocols can be supported. One security protocol to fit all needs would be simpler, but network configurations from small office networks to large-scale Internet content providers do not share the same security requirements. Customers need to have choices of how to integrate new security technology, such as dynamic passwords or public-key cryptography, into their computing environment.

Windows 2000 is designed to support multiple security protocols, an essential element for today's distributed computing environment. Using general-purpose Win32® security APIs, the operating system isolates supported applications from the details of different security protocols available. Higher-level application interfaces provided by Authenticated RPC and DCOM provide abstractions based on interface parameters to use security services.

The Windows 2000 security infrastructure supports these primary security protocols:

- Windows NT LAN Manager (NTLM) authentication protocol is used by Windows NT 4.0 and previous versions of Windows NT. NTLM will continue to be supported and used for pass-through network authentication, remote file access, and authenticated RPC connections to earlier versions of Windows NT.

  The Kerberos Version 5 authentication protocol replaces NTLM as the primary security protocol for access to resources within or across Windows 2000 domains. The Kerberos authentication protocol is a mature industry standard that has advantages for Windows network authentication. Some of the benefits of Kerberos protocol are mutual authentication of both client and server, reduced server load during connection establishment, and support for delegation of authorization from clients to servers through the use of proxy mechanisms.
- Distributed Password Authentication (DPA) is the shared secret authentication protocol used by some of the largest Internet membership organizations, such as MSN and CompuServe. This authentication protocol is part of Microsoft Commercial Internet System (MCIS) services and is specifically designed for users to use the same Internet membership password to connect to any number of Internet sites that are part of the same membership organization. The Internet content servers use the MCIS authentication service as a back end Internet service, and users can connect to multiple sites without reentering their passwords.
- Public-key-based protocols provide privacy and reliability over the Internet. SSL is the *de facto* standard today for connections between Internet browsers and Internet information servers. (A forthcoming IETF standard protocol definition based on SSL3 is currently known as the Transport Layer Security Protocol, or TLS.) These protocols, which use public-key certificates to authenticate clients and servers depend on a public-key infrastructure for widespread use. Windows NT 4.0 provides secure channel security services that implement the SSL/PCT protocols. Windows 2000 security has more enhanced feature support for public-key protocols, which are described later in this paper.

Enterprise security depends on having the flexibility to use the right security mechanisms when necessary. Enterprise computing will continue to depend on a wide range of network services provided by remote file and print servers, business application and data servers, and data warehouse and transaction processing environments. Support for multiple network security protocols allows Windows 2000 Professional and Windows 2000 Server to host a variety of network services in addition to Internet-based technologies.

The following diagram shows the architecture support for multiple security protocols implemented in Windows 2000 using the Security Support Provider Interface (SSPI).

Architecture for Multiple Authentication Services

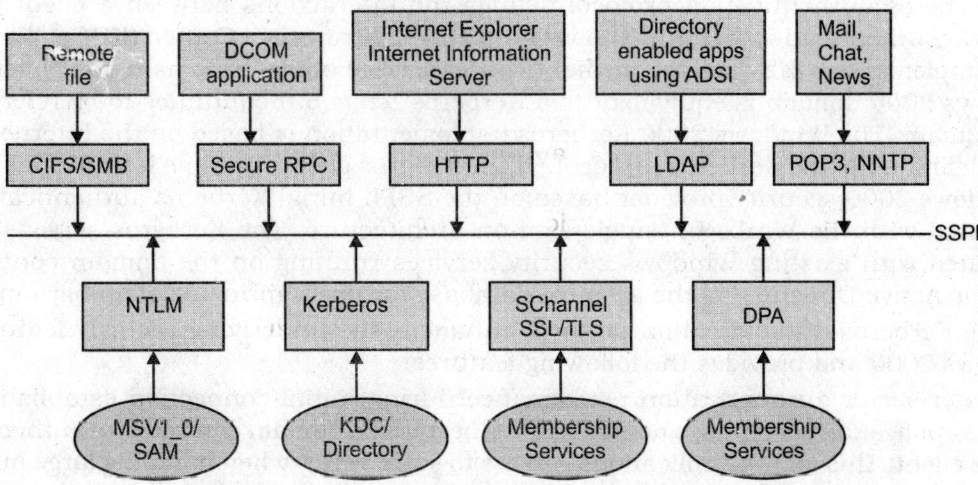

FIG. 9.4 Architecture for Multiple Authentication Services

The Security Support Provider Interface is a Win32 system API used by many applications and system services—for example, Internet Explorer (IE) and Internet Information Server (IIS)—to isolate application-level protocols from security protocols used for network authentication. Security providers use different credentials to authenticate the user, either shared-secret or public-key certificates. The security protocols interact with different authentication services and account information stores.

- NTLM security provider uses the MSV1_0 authentication service and NetLogon service on a domain controller for client authentication and authorization information.
- Kerberos security provider connects to an online Key Distribution Center (KDC) and the Active Directory account store for session tickets.
- DPA uses the MCIS security services for membership authentication and server-specific access information.
- Secure channel services are based on public-key certificates issued by trusted Certificate Authorities; they do not require an online authentication server.

9.3 SECURITY SUPPORT PROVIDER INTERFACE

The Windows security APIs for network authentication are defined by the Security Support Provider Interface (SSPI) documented in the Platform SDK. The SSPI communicates with a Win32 API based on the Generic Security Service Application Program Interface (GSS-API) and provides similar interface abstraction for security context management.[1] Windows 2000 applications and services use SSPI to isolate application-level protocols from the details of network security protocols. Windows 2000 supports the SSPI interface to reduce application-level code needed to support multiple authentication protocols. SSPI provides a generic abstraction to support multiple authentication mechanisms based on shared-secret or public-key protocols. Applications using integrated Windows 2000 security take advantage of the modularity provided by SSPI by calling the SSPI routines directory or by using the higher-level network connection management protocols provided by authenticated RPC or DCOM.

9.4 KERBEROS AUTHENTICATION PROTOCOL

The Kerberos authentication protocol defines the interactions between a client and a network Authentication Service known as a Key Distribution Center (KDC). Windows 2000 implements a KDC as the authentication service on each domain controller. The Windows 2000 domain is equivalent to a Kerberos *realm* but continues to be referred to as a domain. The Windows 2000 Kerberos implementation is based on the Internet RFC 1510 definition of the Kerberos protocol.[2] The Kerberos client run time is implemented as a Windows 2000 security provider based on the SSPI. Initial Kerberos authentication is integrated with the WinLogon single-sign-on architecture. The Kerberos server (KDC), integrated with existing Windows security services running on the domain controller, uses the Active Directory as the account database for users (principals) and groups.

The Kerberos authentication protocol enhances the underlying security features of Windows 2000 and provides the following features:

- Faster server authentication performance during initial connection establishment. The application server does not have to connect to the domain controller to authenticate the client. This allows applications servers to scale better when handling large number of client connection requests.
- Delegation of authentication for multitier client/server application architectures. When a client connects to a server, the server impersonates the client on that system. But if the server needs to make a network connection to another back-end server to complete the client transaction, the Kerberos protocol allows delegation of authentication for the first server to connect on behalf of the client to another server. The delegation allows the second server to also impersonate the client.
- Transitive trust relationships for interdomain authentication. Users can authenticate to domains anywhere in the domain tree because the authentication services (KDCs) in each domain trust tickets issued by other KDCs in the tree. Transitive trust simplifies domain management for large networks with multiple domains.

The Kerberos version 5 authentication protocol defined in RFC 1510 has gone through a wide industry review and is well known in security interest groups.

9.4.1 Kerberos Background

Kerberos is a shared-secret authentication protocol because the user and the KDC both know the user's password or, in the case of the KDC, the one-way encrypted password. The Kerberos protocol defines a series of exchanges between clients, the KDC, and servers to obtain and use Kerberos tickets. When a user initiates a logon to Windows, the Kerberos SSP obtains an initial Kerberos ticket (TGT) based on an encrypted hash of the user's password. Windows 2000 stores the TGT in a ticket cache on the workstation associated with the user's logon context. When a client program attempts to access a network service, the Kerberos run-time checks the ticket cache for a valid session ticket to the server. If a ticket is not available, the TGT is sent in a request to the KDC for a session ticket that allows access to the server.

The session ticket is added to the ticket cache and may be reused for future connections to the same server until the ticket expires. The ticket expiration period is defined by domain security policy and is usually set for about eight hours. If the session ticket expires during the middle of an active session, the Kerberos security provider returns appropriate error values that allow the client and server to refresh the ticket, generate a new session key, and resume the connection.

The following diagram shows the relationship between the client, the KDC, and the application server using the Kerberos authentication protocol.

9.5 INTERNET SECURITY FOR WINDOWS 2000

Microsoft is developing a public-key security infrastructure to integrate public-key security with Windows 2000 security. Public-key cryptography is the security technology that enables strong security for enterprise and Internet communications. Microsoft Internet security technologies include a Certificate Server, a secure channel security provider that implements SSL/TLS protocols, the SET secure payment protocol for credit card transactions, and CryptoAPI components for certificate management and administration.

The components of the Microsoft public-key security infrastructure are shown below.

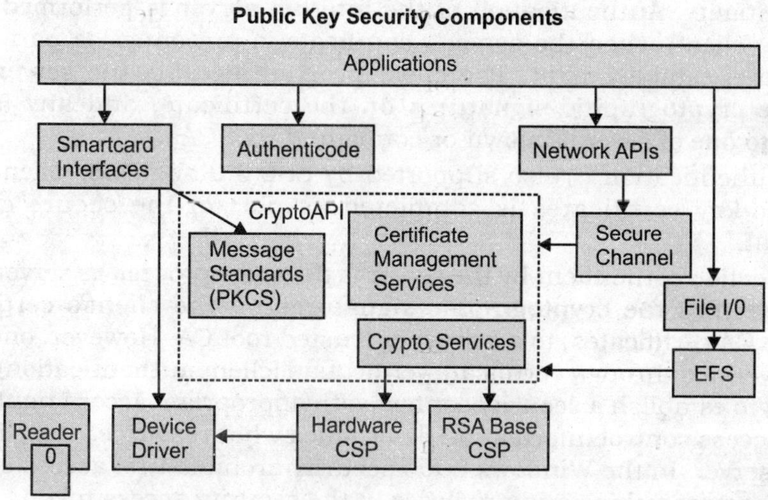

FIG. 9.5 Microsoft public-key security components

The Microsoft Internet security infrastructure is based on industry standards for public-key security, including support for RSA Public-key Cipher, X.509 certificate formats, and PKCS standards.

The Windows NT 4.0 release provided the first components for using public-key security, including:

- CryptoAPI, with programmer support for key generation and exchange, digital signatures, and data encryption, using provider architecture to support installable Cryptographic Service Providers.
- Crypto API support of X509 certificates and PKCS, which was released in Service Pack 3 for Windows NT 4.0 and is used by Internet Explorer 4.0 and Windows 2000.
- Secure Channel implementation of the Secure Sockets Layer (SSL) version 2.0, version 3.0 client-side support, and Private Communications Technology (PCT) version 1.0 public-key security protocols.
- Authenticode™ an industry-standard solution, using digital signatures to verify the integrity of software downloaded from the Internet and identification of the software publisher.

The Microsoft Internet security infrastructure builds on these components and provides additional functionality to support public-key security for Windows platforms, including Windows 2000. Many of the Internet security components are used by Microsoft Internet

Explorer and Internet Information Server. The new features of the Microsoft Internet security infrastructure for Windows 2000 Distributed Security Services include:

- Client Authentication with SSL 3.0 based on public-key certificates.
- Certificate Server for issuing certificates to Windows 2000 domain accounts.

Windows 2000 security uses Internet standards for public-key security with features built into the operating system.

9.6 CLIENT AUTHENTICATION WITH SSL 3.0

Secure Socket Layer and Transport Layer Security are public-key-based security protocols implemented by the Secure Channel (Schannel) security provider. These security protocols are used by Internet browsers and servers for mutual authentication, message integrity, and confidentiality. Authentication of the Internet server is performed by the Internet Explorer (the client) when the server's certificate is presented as part of the SSL/TLS secure channel establishment. The client program accepts the server's certificate by verifying the cryptographic signatures on the certificate, and any intermediate CA certificates, to one of several known or configured root CAs.

Client authentication is also supported by SSL 3.0 and TLS. Client authentication using public-key certificates is completed as part of the secure channel session establishment.

Authentication of the client by the server is the same process as server authentication. The server verifies the cryptographic signatures on the client's certificate, and any intermediate CA certificates, to a known or trusted root CA. However, once the identity of the client is verified through certificate verification (client authentication), the application server needs to establish a security context with appropriate access rights defined for the client. The access control information determines what resources the client is allowed to use on this server. In the Windows 2000 security architecture, access control is defined by the group memberships and privileges in the security access token.

Public-key client authentication uses the information in the client's certificate to map to local access control information. This mapping determines what *authorization* the client has to access resources on the server system. Initial support for client authentication by the Microsoft Internet Information Server is available by managing an authorization database to map certificate subject or issuer information to existing Windows 2000 accounts. The authorization database can be as simple or complicated as needed to meet the application requirements.

Windows 2000 provides broader support for client authentication by implementing a security service that uses the Active Directory to map certificate information to existing Windows accounts. The mapping can be performed using a search of the certificate subject name in the Windows directory or by searching for directory properties that identify the client certificate.

Windows 2000 support for client authentication integrates public-key certificates with the Windows 2000 security architecture. No separate database is required to define the access rights associated with public-key certificates. The access control information is maintained by the group membership stored in the Windows directory. Common Windows Directory Service administration tools are used for granting access rights by adding Windows users to groups.

9.6.1 Authentication of External Users

Support for public-key certificate authentication in Windows 2000 allows client applications to connect to secure services on behalf of users who do not have a Windows 2000 domain account. Users who are authenticated based on a public-key certificate issued by a trusted Certificate Authority can be granted access to Windows 2000 resources. The Directory Service administration tools allow administrators, or delegated authorities, to associate one external user or more to an existing Windows 2000 account for access control. The Subject name in the X.509 Version 3 certificate is used to identify the external user that is associated with the account.

Businesses can share information securely to selected individuals from other organizations without having to create many individual Windows 2000 accounts. Many-to-one mapping of certificates to Windows 2000 user objects provides for strong authentication based on public-key certificates and common access control permissions. Client authentication of external users still requires the system administrator to configure the Certificate Authority for the external user's certificates as a trusted CA. This prevents someone with a certificate issued by an unknown authority from authenticating to the system as someone else.

9.6.2 Microsoft Certificate Server

The Microsoft Certificate Server, which is included with Windows 2000 and IIS 4.0, provides customizable services for issuing and managing certificates for applications using public-key cryptography. The Certificate Server can perform a central role in the management of such a system to provide secure communications across the Internet, corporate intranets, and other nonsecure networks. The Microsoft Certificate Server is customizable to support the application requirements of different organizations.

The Certificate Server receives requests for new certificates over transports such as RPC, HTTP, or e-mail. Each request is checked against custom or site-specific policies, sets optional properties of the certificate to be issued, and issues the certificate. It also allows administrators to add elements to a certificate revocation list (CRL), and publish a signed CRL on a regular basis. Programmable interfaces are included for use by developers to create support for additional transports, policies, certificate properties, and formats.

The policy module for the Certificate Server uses network authentication of certificate requests to issue certificates to users with Windows 2000 domain accounts. The policy module may be customized to meet the needs of the issuing organization. Certificate Server generates certificates in a standard X.509 format. Certificates in the X.509 format are commonly used to authenticate servers and clients involved in secure communications, using either the TLS or SSL protocols. The following sections describe uses of and some key features of Certificate Server.

On a corporate intranet or on the Internet, servers such as the Microsoft Internet Information Server can perform client authentication for secure communications using certificates generated by the Certificate Server. Certificate Server can also generate server certificates used by IIS and other Web servers to provide server authentication to assure clients (browsers) that they are communicating with the intended entity.

9.6.3 CryptoAPI

Windows NT 4.0 provided the low-level cryptography support and modular Cryptographic Service Providers in CryptoAPI. Windows 2000 benefits from the introduction of CryptoAPI Certificate Management to support public-key security.

Major new features of CryptoAPI include:

- Support for X.509 version 3 certificates and X.509 version 2.0 CRLs through common encoding/decoding functions, certificate parse, and verification.
- Support for PKCS #10 certificate requests and PKCS #7 for signed and enveloped data.
- Adding and retrieving certificates and CRLs from certificate stores, locating certificates by attributes and associating private keys.
- Digital signature and verification and data encryption support using higher-level functions available to applications in HTML, Java, Visual Basic® Scripting Edition (VBScript), and C/C++.

CryptoAPI features are used by Windows 2000 operating system components, such as the Software Publisher Trust Provider for Authenticode verification. Other applications and system services use CryptoAPI version 2.0 to provide the common functionality to enable public-key security technology.

9.6.4 Interbusiness Access: Distributed Partners

Internet-based enterprises are already doing business with customers and partners over the Internet. Resellers, suppliers, distributors, and anyone who is part of the extended business may connect to corporate intranets and access important company information. Employees and representatives in the field increasingly use local access to public networks, and then connect to remote corporate information sources. Windows NT security is evolving to support the changing needs of distributed computing over the Internet.

Interbusiness distributed computing is not limited to a single architecture, and the security technology should not limit business to a single way of accessing information. Many approaches are available as security technology rapidly changes. Windows 2000 integrates support for the security protocols and user models that fit application or business needs. More importantly, Windows 2000 provides a migration from the enterprise security in use today, with the opportunity to fully utilize Internet public-key security as the infrastructure matures.

Here are some of the options Windows 2000 security provides for managing and supporting interbusiness relationships:

- An initial approach widely used today is creating user accounts for business partners to access corporate information services. Integrating Windows 2000 security with the Active Directory makes management of these special accounts much easier. Organizational units in the directory can be used to group related accounts by partner, supplier, or other business relationship. Administration of these accounts can be delegated to the people in the organization who manage these partner relations. Virtual Private Networks are established between organizations to encrypt network traffic carried over the public network. Using this approach, business partners can use remote access services to get to corporate information in the same way as any other remote employee. Access to databases or information repositories can be controlled with Windows 2000 access control.
- Domain-trust relationships are another tool for establishing cross-business relationships. The Active Directory provides much more flexibility to manage a tree of hierarchical domains. With Windows 2000 domain names integrated with DNS naming, Internet routing of information between two domains is easy to configure. If the business relationship requires it, domain trust can be used as one way to configure client/ server applications that also have the privacy and integrity features necessary to

communicate over the Internet. Users can use either Kerberos or public-key authentication protocols to access shared resources in remote domains.

- Organizations can use the Microsoft Internet security infrastructure to solve Internet security problems. Companies can issue public-key certificates to specific partners who need to access specific information resources. Instead of creating a user account or defining a domain trust relationship, certificates can be used as a way of providing user identification and authorization. Public-key certificates—and the infrastructure required to support issuing certificates and verifying certificate revocation—are the most effective ways to support business-to-business relationships over the Internet. Windows 2000 supports X.509 version 3 certificates issued by any certificate issuing system. System administrators on Windows 2000 define which Certificate Authorities are trusted. They can also associate external users authenticated by public-key certificates to Windows 2000 user accounts to define the access permissions associated with those users.

9.6.5 Enterprise and Internet Single Sign-on

Windows 2000 manages the user's network security credentials transparently after a single successful sign-on. The user is not concerned about whether a connection to a network server uses NTLM, Kerberos or a public-key-based security protocol. From the users' perspective, they have signed on to the system and now have access to a wide variety of network services.

Within the enterprise, access to resources is determined by the rights granted to users' accounts or by group memberships. Across the Internet, a user's access is based on their identity proven by a private-key signature operation and the corresponding public-key certificate. All of the security protocols rely on some form of user credentials, which are presented to a server when a connection is established. Windows 2000 manages these user credentials and automatically uses the appropriate set of credentials, based on the security protocol involved.

The Windows 2000 Active Directory supports multiple security credentials as part of the secure portion of user account information. These credentials are used for enterprise authentication services that use the domain controller for online user authentication. Advanced application servers can support integrated Windows 2000 authentication by using the Security Service Provider Interface for network authentication.

NTLM Credentials

The NTLM authentication protocol is used by Windows 2000 clients to connect to servers running previous versions of Windows NT. For example, NTLM authentication is used to connect to a remote file share on a Windows NT 4.0 server or to connect a Windows NT 4.0 client to a Windows 2000 file share. NTLM credentials consist of the domain name, user name, and encrypted password entered once during the initial sign-on.

The security services on a domain controller manage a secure copy of the NTLM user credentials in the Active Directory to use for NTLM authentication. A Windows 2000 client manages the NTLM credentials entered at system sign-on on the client side to use when the client connects to Windows NT 4.0 servers using NTLM authentication. Support for NTLM credentials in the Windows 2000 security is the same as for Windows NT 4.0 for compatibility.

Kerberos Credentials

The primary authentication protocol for the Windows 2000 domain is Kerberos authentication. Kerberos credentials consist of the domain and user name (which could be in the form of Internet friendly names, such as BobbyB@microsoft.com), and the Kerberos-style encrypted password. When the user signs on to the system, Windows 2000 obtains one or more Kerberos tickets to connect to network services. The Kerberos tickets represent a user's network credentials in Kerberos-based authentication.

Windows 2000 automatically manages the Kerberos ticket cache for connections to all network services. Tickets have an expiration time and occasionally need to be renewed. Ticket expiration and renewal are handled by the Kerberos security provider and associated application services. Most services, such as the file system Redirector, automatically keep session tickets up-to-date. Regular ticket renewal gives added session security by changing the session keys periodically.

Private/Public-Key Pairs and Certificates

Internet credentials in the form of private/public-key pairs and certificates are managed by the user. The Active Directory is used to publish public-key certificates for users, and standard directory access protocols are used to locate them. Private keys and certificates issued to end users are kept in secure storage, either on the local system or on a smart card. The secure storage is provided with the Internet security technologies and is known as a Protected Store.

The implementation of the Protected Store is based on the CryptoAPI architecture for Windows NT. CryptoAPI provides key management functionality and other cryptographic capabilities for building a secure store, with certificates kept in a Certificate Store. The Windows 2000 implementation of public-key-based security protocols uses keys and certificates accessed from the Protected Store and Certificate Store as user credentials for accessing Internet-based servers. In many cases, user-defined properties of certificates in the Certificate Store allow the security protocols to automatically select and use the correct certificate and signature key. Advances in Internet security protocols (SSL3/TLS) allow a server to request specific credentials from the client that are used automatically from the Certificate Store if they are available.

The information in the Protected Store and Certificate Store is available to roaming users because they are implemented securely as part of the user's profile. When a user initially logs on to the Windows client, the user's profile information is copied to that computer. If the user gets new keys and certificates during that session, the user profile is updated to the central server when the user logs off.

Seamless Transition

The transition from the NTLM authentication used in Windows NT 4.0 (and previous versions of Windows NT) to Kerberos domain authentication will be very smooth. Windows 2000 services supports client or server connections using either security protocol. Security negotiation, either by the SSPI layer or the application protocol, provides another option to select the best match from available security protocol options.

The transition from enterprise-based services using Kerberos authentication to Internet-based services using public-key authentication is completely transparent to the user. Windows 2000 support for multiple user credentials makes it possible to use secret-key authentication technology for enterprise application services with very high performance and public-key security technology when connecting to Internet-based servers. Most application protocols, such as LDAP, HTTP/HTTPS, or RPC, support authentication, and they are designed to support multiple authentication services and select those services during connection establishment.

Rather than relying on a single authentication technology and a single authentication protocol, Windows 2000 uses multiple protocols as needed to fit the application and user-community requirements for secure network computing.

Providing a Smooth Migration to Next-Generation Domains

Migrating from a Windows NT 4.0 environment to the Windows 2000 domain is easy because of backward compatibility with existing Windows security and account replication protocols. A smooth migration is available because Windows 2000 has the following interoperability features:

- A Windows 2000 domain controller can play the role of a Windows NT 4.0 BDC and receive domain account replication from an existing Windows NT 4.0 PDC.
- Windows NT 4.0 workstations can send network authentication requests using the NTLM authentication protocol to a Windows 2000 domain controller, acting as a BDC in the Windows NT 4.0 domain.
- Windows 2000 domain controllers can establish trust relationships with Windows NT 4.0 domains and support pass-through authentication between domains. This means that not all domains in an enterprise are required to upgrade to the Windows 2000 domain security at the same time.
- Windows 2000 domain controllers can eventually replace Windows NT 4.0 domain controllers in a gradual upgrade of Windows NT 4.0 BDCs to Windows 2000 domain controllers. Windows NT 4.0 account management tools are used on the primary domain controller as long as the PDC is running Windows NT 4.0. Eventually, all domain controllers can be upgraded to use the Active Directory for account management and multimaster account replication.

Windows 2000 support for multiple authentication protocols means that from a single domain logon at the desktop, users can access Windows 2000 services anywhere in a mixed domain environment, including:

- A Windows 2000 server in the logon, or home domain, using Kerberos session tickets issued by the Key Distribution Center (KDC) on the domain controller.
- A Windows 2000 server in a trusted domain, using a Kerberos referral to the KDC in the trusted domain, to issue a session ticket to the remote server.
- Or, a Windows NT 4.0 server in a trusted domain using NTLM pass-through authentication between the client, the Windows NT 4.0 server, and the trusted domain controller.

Because Windows 2000 continues to support NTLM authentication, Windows NT 4.0 clients who do not use Kerberos authentication can also connect to Windows 2000 application servers.

These interoperability features allow flexibility for organizations to plan and implement a migration strategy to Windows 2000 Servers to better fit their growing business needs.

9.7 LDAP(LINUX LIGHTWEIGHT DIRECTORY ACCESS PROTOCOL)

LDAP is a lightweight directory access protocol package which provides client/server based directory database services which runs on UNIX platforms over TCP/IP. The associated RFC is 1777. Directory service is based on entries which have the following characteristics.

1. Distinguished name (DN)
2. Collection of attributes

1. Types - String such as "mail".
2. Values (one or more) - Example: myself@mycompany.com

The directory hierarchy is organized into geographic, political, or organizational boundaries such as from top to bottom as shown below:

1. countries
2. states or organizations
3. units
4. people, printers, documents

There are three database choices:

1. LDBM - High performance disk based database
2. SHELL - Database interface to UNIX commands.
3. PASSWD - Password file database.

LDAP capabilities are:

- Can serve multiple databases
- Provides access control
- Can replicate the database

SLAPD - LDAP database Server

9.7.1 Network Information Services

When managing a large number of computers it is convenient to store configuration data in a central location rather than maintain separate files on each machine. Thus, DNS quickly replaced large /etc/hostsfiles on the Internet, and a number of more general nameservices were developed to serve more local needs. Sun's *YP/ NIS* is probably the best known, but *Hesiod[1]* does a similar job and protocol suites such as *XNS* and *NetBIOS* also include nameservices of one sort or another.

The more general nameservices support a number of 'maps', each of which translates from keys to values. A NIS serving Unix machines has maps such as passwd, group, automount, services, and rpc. Often there are two or more key fields in a map: the passwd map for example has to service lookups both by username and by uid. Data is accessed with the same library calls that originally handled local files, so most application programs are not aware that the underlying storage has changed.

A network using a NIS relies on it completely for normal operation so maintaining its integrity is vital. Most NIS systems in use today have very little protection against active attackers: they depend on the IP address of the server being widely known and assume that nobody has taken over its address. This is obviously insecure in a world where TCP connections can be hijacked at a distance, so system designers are turning to cryptography to provide assurance that good data is being used.

9.7.2 Network Authentication Service

Authenticating users is one of the more visible aspects of computer security. People are used to providing usernames and passwords when they want to use a machine or other resource.

One approach to authentication in a network is simply to use the NIS to access conventional Unix password hashes and to do the validation locally on the desktop machine. This scheme is commonly used with *YP/ NIS* but it suffers from having to make all password hashes available to all desktop machines. There are shadow-password schemes that prevent 'ordinary' users from getting hold of the hash data, but these are fairly easy to bypass when used with a NIS. With access to a collection of password hashes, a cracker can mount a dictionary attack with a good chance of success so it would be better to keep the hashes away from client machines entirely.

Another approach is to use a special-purpose authentication protocol such as RADIUS or TACACS+. These were originally developed to control access to modem banks, but are equally usable with desktop computers. Kerberos is another way to do authentication without exposing secrets: this also has the benefit of providing a complete single-sign-on environment if it is fully implemented. Several other protocols have been pressed into service for authentication over the years: SMB allows authentication against an NT domain, and I have even heard of POP being used in this way! It comes as no surprise to find that LDAP is now used an authentication protocol as well.

Ideally, a network authentication service should provide reliable answers, be fast, and should work without exposing any information about the password. This is harder than it first looks, as the client does not want to expose the password to the server in case the server turns out to be compromized. Similarly, the server should not expose any hash data to the client. Network snoopers should not be able to gather any data that can be replayed later to gain unauthorised access. Each protocol tackles these problems in different ways, with varying degrees of success. LDAP has a number of authentication and security options which can provide very good security if used correctly, though as ever there is a trade-off between security and ease of management.

9.7.3 LDAP Basics

LDAP[6] is a directory-access protocol derived from X.500. It works with a tree structure where each node or object in the tree contains a set of attribute-value data. Each object belongs to one or more objectclasses, which define the mandatory and optional attributes. The original application of both X.500 and LDAP was to provide a 'white pages' directory service, where most objects in the tree represented people and the tree had a geographic or organizational structure.

9.7.4 Storing NIS Data in an LDAP Directory

RFC2307[2] describes a schema for storing NIS data in an LDAP directory. Mappings are provided for all the common databases: passwd, group, hosts, shadow, services, netgroup, protocol, ethers, etc. The RFC describes objectclasses and attributes, but does not mandate a particular tree structure. A common practice is to use a structure like this:

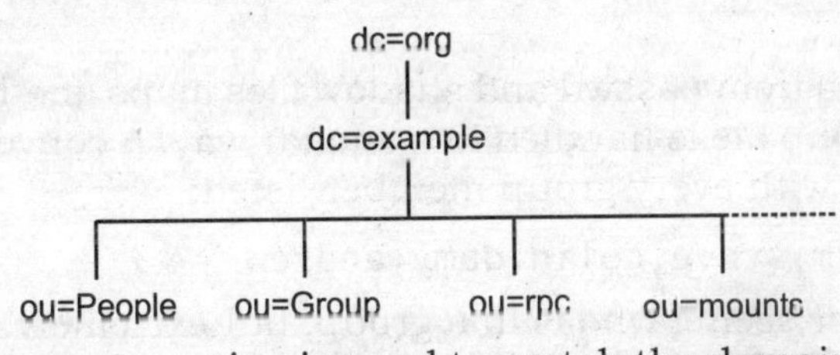

Here, domain-component naming is used to match the domain name *example.org* and separate subtrees are provided for each type of information. Data from the passwd file goes into ou=People rather than ou=passwd because it is assumed that each person has one primary account and that other information will be added to the same entry (phone number, e-mail address, etc).

Passwd data

Typical data in an account entry looks like this:

```
dn:  uid=andrew,ou=People,dc=example,dc=org
uid:  andrew
cn:  Andrew  Findlay
objectClass:  account
objectClass:  posixAccount
objectClass:  top
objectClass:  shadowAccount
shadowLastChange:  11296
shadowMax:  99999
shadowWarning:  7
loginShell:  /bin/bash
uidNumber:  500
gidNumber:  500
homeDirectory:  /home/andrew
gecos:  Andrew  Findlay
userPassword:  {crypt}$1$uQSw.ohy$XuiRSCqOkp...
```

The first line sets the Distinguished Name - the unique name identifying this entry in the tree. *uid* is used as the naming attribute: this refers to the User ID that we would normally call 'username' in a Unix context. There is one aspect of*uid* to be particularly aware of: it is matched in a case-insensitive way so it is not possible to represent distinct accounts 'fred' and 'FRED'.

Notice that the *uid* attribute occurs again inside the entry, along with *uidNumber* and *gidNumber*: there must be exactly one of each of these attributes and they are used when resolving calls to *getpwnam()* etc.

The full name of the account owner appears twice: first as *cn* (Common Name) which is an attribute that can take multiple values and is much used by white-pages lookups. The extra copy in the *gecos* attribute allows control over the exact string returned to *getpwnam()* calls.

The *objectClass* attributes refer to aspects of the RFC2307 schema. They set the list of mandatory and optional attributes for the entry, and are also useful in searches. A call of the form *getpwnam("andrew")* is likely to result in an LDAP search of the form (uid=andrew) AND (objectClass=posixAccount)

Group data

As we have seen, data from passwd and shadow files maps line-by-line into the account entry. Data from the group file is handled in a similar way. A conventional Unix group file has one group per line, with every group member listed:

```
ldapbods:x:389:tim,steve,colin,damy,andrew
```

This is convenient for seeing who is in a group, but extremely inefficient for the most common operation: listing the groups that a particular user is a member of at login time. The usual algorithm is to enumerate every entry in the group map looking for occurrences of the username.

RFC2307 defines group membership using the*memberUid* attribute in a *posixGroup* object. Because*memberUid* is a multi-valued attribute, the group map can now be searched very efficiently. The example group entry above would look like this in the directory:

```
dn:  cn=ldapbods,ou=Group,dc=example,dc=org
objectClass:  posixGroup
objectClass:  top
cn:  ldapbods
userPassword:  {crypt}x
gidNumber:  389
memberUid:  tim
memberUid:  steve
memberUid:  colin
memberUid:  damy
memberUid:  andrew
```

Now, to find the list of groups that a user belongs to, the search is of the form:

```
(objectClass=posixGroup)  AND  (memberUid=andrew)
```

The LDAP server can be configured with indexes on any attribute, so searches of this form can be made very efficient and large databases can be handled without performance penalties.

9.7.5 Putting it all together

So, much for the theory, but how much of this actually works? To find out, I built a testbed on a portable PC running Red Hat 7.1 and supporting several VMware virtual machines with diverse operating systems. I also connected a Sun Ultra 5 running Solaris 8.

The LDAP server was *slapd* from OpenLDAP 2.0.18 (a security advisory has since been issued requiring upgrade to 2.0.21). The config file is shown in Appendix 1. slapd sends logging data to syslog, and it is a good idea to separate this into a file of its own with a line like this in /etc/syslog.conf:

```
local4.*  /var/log/slapd.log
```

I find it convenient to keep a wide window on screen running 'tail -f' on the logfile while testing.

To create the initial data I used the migration tools from PADL software. These are often included with Linux distributions, but it is worth fetching the latest version from www.padl.com The tools need very little configuration, and will load data from existing files in /etcinto the LDAP directory. Note that the tools create the container entries *ou=People etc* but they do not create the top-level entry (*dc=example,dc=org* in this case) so it must be created by hand before the tools will run. To do this, create a file of the form:

```
dn:  dc=example,dc=org
objectClass:  dcObject
objectClass:  organization
o:  The  Example  Organisation
dc:  example
```

and load it with a command like this:

```
ldapmodify  -a  -H  ldap://localhost/  -r  -c  -x  \
        -D  'cn=Manager,dc=example,dc=org'  -W  \
        -f  <filename>
```

At this stage, the LDAP server can be tested with command-line tools such as *ldapsearch*. Once it is working it is possible to configure systems to use it as a NIS.

The Name Service Switch and nss_ldap

NSS is a facility found in Solaris and Linux which allows new name-resolution code to be installed as shared objects without having to rebuild existing libraries. Some versions of FreeBSD and AIX have a similar concept called the Information Retrieval System, apparently derived from BIND 8 code.

In principle you can install a new name-resolution system just by placing a shared library in the right directory and editing /etc/nsswitch.conf. Solaris and many Linux distributions come with LDAP modules for NSS, though they are often derived from fairly old code and you may need to fetch the latest version from www.padl.com if you want all the security features.

I tested nss_ldap on a virtual machine with a small Red Hat 7.2 installation. The first thing to do is to configure the LDAP parameters in /etc/ldap.conf:

```
# ldap.conf
#
# Location of LDAP server.
# Must be resolvable without using LDAP!
#
host brick.skills-1st.co.uk
#
# The distinguished name of the search base.
#
base dc=example,dc=org
#
```

This simple config is enough to get things going, though we will return to add more security later.

To make the machine use LDAP for passwd and group lookups, edit /etc/ nsswitch.confand change lines as shown:

```
passwd: files ldap
shadow: files ldap
group: files ldap
```

The effect of this is that lookups will use local files first and progress to LDAP for anything not found locally. This allows for system accounts to be kept in /etc files (necessary during bootup) while user accounts are held centrally.

If the name service cache daemon (nscd) is running, it may be necessary to restart it before LDAP lookups work. Similarly, nsswitch.conf is normally only read once per process so don't expect any existing processes to notice until they are restarted (and that includes the shell that you used to set all this up from!)

An easy test is to find the home-directory of a user that is listed in the LDAP store but not in the local password file:

```
echo ~fred
```

The Solaris version of nss_ldap is based on the iPlanet LDAP libraries and it does not use /etc/ldap.conf, preferring instead to use a program called *ldapclient* to set up the parameters. The problem with *ldapclient* is that it depends on a feature that is implemented differently in different LDAP servers so it does not easily work with OpenLDAP. There are a number of mailing-list archives containing discussion of the problem and suggesting various workarounds, though I have not yet made it work myself.

FreeBSD is supposed to be able to use the PADL nss_ldap code, but it requires a rebuild of at least the main C library to make it work. Again, I have not had time to test this, but in principle it should work with the same data that supports Linux.

Authentication with pam_ldap

The usual way of using LDAP for user authentication is to locate the account entry in the directory and then try to 'bind' to the directory as that account, presenting the password or other credentials as received from the user. In the simplest case, the password is presented to the directory server in clear, but there are a number of more secure schemes.

Most Unix-like systems now implement Pluggable Authentication Modules (PAMs) of some form. This allows authentication methods to be changed easily without having to re-compile every program that might need to check a password. PAMs can also implement policy such as time-of-day restrictions and other security requirements. A particularly powerful feature is that PAMs can be stacked, allowing for a choice of authentication methods and perhaps several different policy modules to be active as well. As with NSS, PAM is implemented with shared libraries so re configuration is quite easy.

Red Hat, SuSE, Mandrake, and FreeBSD all provide pam_ldap modules derived from the PADL code. Solaris includes a module based on iPlanet code, though it is quite possible to replace this with the PADL version if required. In all cases, configuration is shared with the equivalent NSS module so little extra work is required.

To use LDAP for authentication is a matter of installing the pam_ldap module and editing the appropriate PAM config: either /etc/pam.confor one or more files in /etc/pam.d. Look for the line describing pam_unix and add pam_ldap at that point. On Red Hat Linux the file is /etc/pam.d/system-authand it looks like this:

```
auth required /lib/security/pam_env.so
auth sufficient /lib/security/pam_unix.so likeauth nullok
auth sufficient /lib/security/pam_ldap.so use_first_pass
auth required /lib/security/pam_deny.so
account required /lib/security/pam_unix.so
account required /lib/security/pam_ldap.so
password required /lib/security/pam_cracklib.so \
retry=3 type=
password sufficient /lib/security/pam_unix.so nullok \
use_authtok md5 shadow
password sufficient /lib/security/pam_ldap.so use_authtok
password required /lib/security/pam_deny.so
session required /lib/security/pam_limits.so
session required /lib/security/pam_unix.so
session optional /lib/security/pam_ldap.so
```

Note that you should not edit the file directly on Red Hat Linux: use the *authconfig* program to maintain it.

Note also that the example config files supplied with the*pam_ldap* source code might give LDAP more control than you want, including the ability to supply an alternate password for*root*.

Solaris uses /etc/pam.conf so the format is slightly different. Here is a snippet describing the authentication requirements for *rlogin*:

```
rlogin auth sufficient \
            /usr/lib/security/$ISA/pam_ldap.so.1
rlogin auth sufficient \
            /usr/lib/security/$ISA/pam_rhosts_auth.so.1
rlogin auth required \
            /usr/lib/security/$ISA/pam_unix.so.1
```

Note the $ISA variable: this is the Instruction Set Architecture, which allows for 32-bit and 64-bit programs to co-exist. Strictly this means that if you replace any PAM modules you should provide both 32-bit and 64-bit versions but so far I have not found any 64-bit programs using authentication This will probably change...

FreeBSD also uses /etc/pam.conf but has a slightly different layout. Here is part of the section for login (and thus rlogin too):

```
login auth sufficient pam_skey.so
login auth requisite pam_cleartext_pass_ok.so
login auth sufficient /usr/local/lib/pam_ldap.so \
try_first_pass
login auth required pam_unix.so try_first_pass
```

Where a PAM password clause has been defined (as in the Red Hat example above), pam_ldap supports changing passwords using LDAP so the user need never know that LDAP is being used rather than local files.

With both nss_ldap and pam_ldap in place on client machines, all user-management can be done through the directory.

9.7.6 Now add Windows: SAMBA PDC

Samba has had the ability to act as an NT-style Primary Domain Controller for some time. Recent releases have added a facility to store the PDC SAM data in an LDAP directory. This is clearly marked as experimental code in Samba 2.2.2 though the Samba-TNG project claims to be more advanced in this area. The two versions have different approaches: mainstream Samba aims to provide Windows-Unix integration, where Samba-TNG aims for a complete PDC implementation that just happens to run on Unix rather than Windows.

If Samba is installed on a machine usingnss_ldap and *pam_ldap* it will of course use them, so networks running SMB with cleartext passwords may not need to do more than that. There are advantages to the PDC model though, so Samba's move to closer integration with LDAP is particularly welcome.

To enable LDAP storage for the PDC, Samba must be configured with the withldapsam option. This completely overrides local storage of SAM data so you cannot use the same binary in both modes.

The LDAP-specific parts of the Samba config file look like this:

```
ldap admin dn = cn=manager,dc=example,dc=org
ldap server = brick.skills-1st.co.uk
ldap suffix = dc=example,dc=org
```

A minimal Samba PDC config is shown in Appendix 2.

Note that the LDAP suffix given here matches the search base used by nss_ldap and pam_ldap. This allows Samba to share the entries that already exist to describe people/accounts. Note also that the admin DN is given: in this case I have used the name of the DSA manager (equivalent to root in directory terms), but it should be possible to use a less powerful ID if ACLs are set up carefully. Samba needs to bind to the directory as a user with write permission so that it can insert new attributes in entries. The password is stored in the secrets database using the command:

```
smbpasswd -w <password>
```

Unfortunately, the password has to be typed on the command line so make sure you clear out any command-history files afterwards!

Samba introduces several new attributes, so it's schema file must be copied from samba-2.2.2/examples/LDAP/samba.schemato the server schema directory and an appropriate include statement has to be added to *slapd.conf*. The schema file supplied tries to re-define *displayName* so this clause must be commented out before restarting the server.

Samba needs to set up two new password hashes and several other attributes for each user. The password hashes cannot be derived from the Unix password data, so be prepared to supply new passwords for each user:

```
smbpasswd -a <username>
```

Unfortunately, this is where the experimental nature of the code starts to show through. Some of the attributes get bad values: in particular, *rid* gets set to zero, and both*smbHome* and *profilePath* get values including %-expansions that are not properly handled when the data is used. These need to be fixed by hand: I used *ldapsearch* to extract the complete attribute set into a file for editing, and*ldapmodify* to put the updated values back.

Before a Windows client can join a domain, it needs a 'trust account' to be set up on the PDC. With the appropriate scripts configured on Samba, this can be largely automated but for test purposes I created Unix-style accounts by hand. These have usernames formed by appending '$' to the Windows 'computername', and are made ready for use with the command:

```
smbpasswd -a -m <computername>
```

With these preparations done, I was able to join an NT client to my domain and log in using the same username and password used with the Unix clients.

There is obviously some work still needed on the LDAP PDC code, but once stable there are some interesting possibilities. One option would be to do away with backup domain controllers altogether, and support distribution and resilience by having replicated LDAP servers each serving the same data to one or more Samba PDCs. I have yet to work out the exact effect of moving a Windows client from one such domain controller to another.

Looking at the problem the other way round, it should be possible to run *nss_ldap* and *pam_ldap* against an Active Directory system. The sample /etc/ldap.conf shipped with the PADL code includes an example configuration for this purpose, though I have not tried it.

9.7.7 Security

With such a concentration of data in the directory, security becomes very important[3]. Anyone who could modify the data could give themselves access to vast numbers of machines at a stroke. Some data needs to be protected from unauthorized viewing: although all passwords are hashed, anyone who can read the hashes can mount a dictionary attack. More subtly, anyone who can hijack a client-server connection can feed bogus data to an individual client, or use the client's privileges to modify server data. All these things can be protected against, and LDAP now has most of the tools needed to do it.

Access Control

Control over who may read what and who may change what is exercised with Access Control Lists (ACLs). This is one of the non-standardized areas of LDAP, and it varies a lot from one server to another. However, all modern servers provide enough control to protect passwords. Here is a simple ACL section from OpenLDAP:

```
access to attrs=userPassword
by self write
by anonymous auth
by * none
access to *
by * read
```

This allows authenticated users to change their own passwords, allows un-authenticated users to authenticate, and prevents all other access to the *userPassword*. It permits read access to everything else. Obviously if Samba PDC data is being stored then similar protection must be extended to*lmPassword* and *ntPassword.*

Client Authentication

The simplest form of client authentication is to bind to the server using a cleartext password. This is the method normally used by *pam_ldap* for checking login passwords. For security, this method should only be used with encrypted connections.

A more secure method is to use one of the SASL authentication mechanisms, such as DIGEST-MD5[4]. This is based on a secret known to both the client and the server, allowing for a simple challenge-response scheme. SASL is also capable of negotiating data encryption to protect subsequent operations.

LDAP also supports encryption and authentication using Transport Layer Security[5]. TLS is closely related to the older SSL scheme, and uses the same certificate-based methods. In its simplest form, TLS provides proof of server identity and protection of data in transit so it is useful where plaintext passwords might be passed across the net. The same mechanism can also be used to prove the identity of the client to the server, where the client has been issued with a suitable X.509 certificate.

It is often more appropriate (and cheaper) to use a private Certification Authority for this purpose rather than getting certificates from a public supplier. This is easy to set up using OpenSSL:

```
openssl genrsa -des3 -out ca.key 2048
openssl req -new -x509 -days 365 -key ca.key -out ca.cert
```

You will be prompted to supply some information. The result of the commands above will be two files: a key file and a certificate file. These are the basis of the Certification Authority.

```
openssl genrsa -out ldap.key 1024
openssl req -new -key ldap.key -out ldap.csr
```

You will again be asked for information to fill in the Certificate Signing Request. The critical item is to put the LDAP server name in when prompted for 'Common Name'. This should be the *exact* name that clients will use when contacting the server. It is best to use a fully-qualified domain name. Note that the *ldap.key* file is not encrypted as it must be read by the LDAP server: protect this file very carefully! Use the CA to sign the certificate:

```
openssl x509 -req -in ldap.csr -out ldap.cert -CA ca.cert \
    -CAkey ca.key -CAcreateserial -days 365
```

The result is a file called *ldap.cert*containing the certificate for the LDAP server. You can examine certificates with the command:

```
openssl x509 -in ldap.cert -text -noout
```

The LDAP server needs copies of its own certificate and key files, and also a copy of the CA certificate. The key file must only be readable by the server process, so make it mode 400 and owned by the owner of the server (OpenLDAP can be run as root, or can be told to set UID to another user after binding the TCP ports). Add lines to *slapd.conf* to tell the server about the new files:

```
TLSCertificateFile   /etc/openldap/keys/server.cert
TLSCertificateKeyFile   /etc/openldap/keys/server.key
TLSCACertificateFile   /etc/openldap/keys/ca.cert
```

When the server is next restarted it will be able to use TLS. It will also support SSL if told to bind to an ldaps:///URL. To make *nss_ldap* and *pam_ldap* use TLS, add a line to /etc/ldap.conf on the client machines:

```
ssl start_tls
```

This provides transport-level encryption of LDAP sessions, thus giving some protection to passwords, etc. However, it does not guard against bogus servers: to get this extra protection it is necessary to give the client machines a copy of the CA certificate that was used to sign the server certificate. Recent versions of *nss_ldap* and *pam_ldap* will verify the server identity if lines of this form are added to /etc/ldap.conf:

```
tls_checkpeer yes
tls_cacertfile  /etc/ssl/ca.cert
```

Now, if a bogus server is encountered the client will immediately close the LDAP connection and deny the attempted login. Note that the reason for the denied login can be hard to find in the client logfiles: it may be necessary to turn on a high level of debugging to see the relevant message! Note also that certificate verification is a fairly recent addition to *nss_ldap* and*pam_ldap* so the versions shipped with most OS distributions do not yet have it.

Samba-2.2.2 has the option to use TLS when accessing LDAP, though it does not yet have code to check server certificates. To enable TLS add this to the global section of the config file:

```
ldap ssl = start tls
```

Other nameservice clients like Automount and AMD do not yet have the ability to use TLS at all, so unfortunately there are still chinks in the armour.

9.7.8 Management Tools

This area is still rather lacking. The PADL migration tools are good for converting and loading the initial data, but once LDAP is in use as the nameservice you will need other tools to create and manage accounts, etc. Scripts can be built around*ldapsearch* and *ldapmodify* from the OpenLDAP distribution, or around one of the Perl modules that support LDAP. Another useful tool is GQ - a general-purpose LDAP browser/modifier.

There is at least one project under way to develop specific management tools for LDAP NIS systems. *ldaputils* is based at SourceForge and eventually plans to have a web-based management interface.

9.7.9 Performance

An LDAP server providing NIS service is likely to handle much more traffic than one just servicing 'white pages' requests so performance is an important consideration in a large network. When sizing servers it is necessary to know how much traffic to expect: Table 1 summarizes the results of some simple tests with both *pam_ldap* and *nss_ldap* configured.

More recent versions of *nss_ldap* make much more efficient use of the server than the older ones. The use of a cache daemon also helps, though it does result in one more long-lived LDAP connection per client machine.

An LDAP server with logging turned off should not have any difficulty in servicing 500 searches per second, but if TLS is in use the cost of connection setup and binding is likely to far outweigh the search load. A large pool of clients will also result in many hundreds of connections being held open, so the server and its supporting operating system must be configured to support a sufficient number of open files.

LDAP servers can be configured to replicate data from a master server, so the read-only query load can be spread across as many machines as necessary. Operations (like password changing) that result in changes to the database get referred back to the master server. The format of /etc/ldap.conf supports multiple servers in the form of a space-separated list, and the servers are tried in turn until one answers or the list is exhausted.

Table 9.1 LDAP load imposed by clients

Client & Operation	Connect	Bind	Search	Connections Held Open Per session
Red Hat 7.1 login	8	8	13*	
Red Hat 7.1 login with nscd	5	5	3	1
Samba connect (not PDC)	5	5	19	1 ❑
Samba connect with nscd (not PDC)	6	6	8	1 ❑
Samba connect with PDC and nscd	2	2	5	0 ❑
Red Hat 7.2 login with nscd	2	4	6	1❑

* Two of these try to return *every* entry from the Group table

❑ Plus one connection per client machine for*nscd.*

SUMMARY

1. The Active Directory provides the store for all domain security policy and account information. The Active Directory, which provides replication and availability of account information to multiple Domain Controllers, is available for remote administration.

2. The Active Directory supports a hierarchical name space for user, group, and computer account information. Accounts can be grouped by organizational units, rather than the flat domain account name space provided by earlier versions of Windows NT.
3. Administrator rights to create and manage user or group accounts can be delegated to the level of organizational units. Access rights can be granted to individual properties on user objects to allow, for example, a specific individual or group to have the right to reset passwords, but not to modify other account information.
4. Active Directory replication allows account updates at any domain controller, not just the primary domain controller (PDC). Multiple master replicas of the Active Directory at other domain controllers, which used to be known as backup domain controllers (BDCs), are updated and synchronized automatically.
5. Windows 2000 employs a new domain model that uses the Active Directory to support a multilevel hierarchy tree of domains. Management of trust relationships between domains is simplified through tree-wide transitive trust throughout the domain tree.
6. Windows security includes new authentication based on Internet standard security protocols, including Kerberos Version 5 and Transport Layer Security (TLS) for distributed security protocols, in addition to supporting Windows NT LAN Manager authentication protocols for compatibility.
7. The implementation of secure channel security protocols (SSL 3.0/TLS) supports strong client authentication by mapping user credentials in the form of public-key certificates to existing Windows NT accounts. Common administration tools are used to manage account information and access control, whether using shared secret authentication or public-key security.
8. Windows 2000 supports the optional use of smart cards for interactive logon, in addition to passwords. Smart cards support cryptography and secure storage for private keys and certificates, enabling strong authentication from the desktop to the domain.
9. Accounts for users, groups, and machines can be organized into directory containers called organizational units (OUs). A domain can have any number of OUs organized in a tree-structured name space. Businesses can organize the name space for account information to represent the departments and organizations in the company. User accounts, as well as OUs, are directory objects that can easily be renamed within the domain tree as the organization changes.
10. The Active Directory supports a much larger number of user objects (more than 1 million objects) with better performance than the registry. Individual domain size is no longer limited by performance of the security account repository. A tree of connected domains can support much larger, complex organizational structures.

SHORT QUESTIONS WITH ANSWERS

1. Verify whether clhcllent is running.

Ans. `paul@debian5 : ~$ ps fax | grep dhclient`

2. Display your current ip address(es).

Ans.
```
paul@debian5:~$ /sbin/ifconfig | grep 'inet'
   inet addr:192.168.1.31 Bcast : 192.168.1.255 Mask:255.255.255.0
   inet addr:127.0.0.1 Mask:255.0.0.0
```

3. Display the configuration file where this ip address is defined.

Ans.
```
Ubuntu/Debian:  cat/etc/network/interfaces
Redhat/Fedora:  cat/etc/sysconfig/network-scripts/ifcfg-eth*
```

4. Follow the nlc configuration in the book to change your ip address from dhcp client to fixed. Keep the same ip address to avoid conflicts!

Ans.
```
Ubuntu/Debian:
ifdown eth0
```

```
vi /etc/network/interfaces
ifup eth0
Redhat/Fedora:
ifdown echo
vi /etc/sysconfig/network-scripts/ifcfg-eth0
ifup eth0
```

5. Did you also configure the correct gateway in the previous question ? If not, then do this now.

6. Verify that you have a gateway.

```
paul@debian5:~$ /sbin/route
Kernel IP routing table
Destination   Gateway      Genmask         Flags  Metric  Ref  Use  Iface
192.168.1.0   *            255.255.255.0   U      0       0    0    eth0
default       192.168.1.1  0.0.0.0         UG     0       0    0    eth0
```

7. Verify that you can connect to the gateway, that it is alive.

Ans.
```
paul@debian5:~$ ping -c3 192.168.1.1
PING 192.168.1.1 (192.168.1.1) 56(84) bytes of data.
64 bytes from 192.168.1.1: icmp_seq-1 ttl=254 time=2.28 ms
64 bytes from 192.168.1.1: icmp_seq=2 ttl=254 time=2.94 ms
64 bytes from 192.168.1.1: icmp_seq=3 ttl=254 time=2.34 ms
---192.168.1.1 ping statistics---
3 packets transmitted, 3 received, 0% packet loss, time 2008ms
rtt min/avg/max/mdev = 2.283/2.524/2.941/0.296 ms
```

8. Change the last two digits of your mac address.

Ans.
```
[root@rhel6 ~]# ifconfig eth0 hw ether 0.8:00:27:ab:67:XX
```

9. Which ports are used by http, pop3, ssh, telnet, nntp and ftp ?

Ans.
```
root@rhel6 ~# grep ^'http '/etc/services
http      80/tcp     www www-http    #WorldWideWeb HTTP
http      80/udp     www www-http    #HyperText Transfer Protocol
root@rhel16   -# grep '^smtp' /etc/services
smtp        25/tcp     mail
smtp        25/udp     mail
root@rhel6    ~# grep ^'ssh' /etc/services
ssh         22/tcp     # The Secure Shell (SSH) Protocol
ssh         22/udp     # The Secure Shell (SSH) Protocol
root@rhel16   ~# grep  ^'telnet' /etc/services
telnet      23/tcp
telnet      23/udp
root@rehel6   ~# grep ^'nntp ' /etc/services
nntp        119/tcp  readnews untp #USENET News Transfer Protocol
nntp        119/udp  readnews untp #USENET News Transfer Protocol
root@rhel6 ~# grep ^ftp' /etc/services
ftp         21/tcp
ftp         21/udp     fsp fspd
```

Note that **sctp** was ommitted from the screenshot.

10. Explain why e-mail and websites are sent over tcp and not udp.

Ans. Because tcp is reliable and udp is not.

11. Display the hostname of your computer.

Ans.
```
paul@debian5:-$ hostname
debians
```

12. Which ip-addresses did your computer recently have contact with?

Ans.
```
root@rhel6 ~# arp -a
? (192.168.1.1) at 00:02:cf:aa:68:f0 [ether] on eth2
? (192.168.1.30) at 00:26:bb:12:7a:5e [ether] on eth2
? (192.168.1.31) at 08:00:27:8e:8a:a8 [ether] on eth2
```

13. Install wireshark on your computer (not inside a virtual machine).

Ans.
```
Debian/Ubuntu : aptitude install wireshark
Red Hat/Mandriva/Fedora: yum install wireshark
```

14. Start a ping between your computer and another computer.

Ans.
```
ping $ip_address
```

15. Start sniffing the network.

Ans.
```
(sudo) wireshark
select an interface (probably eth0)
```

16. Display only the ping echo's in the top pane using a filter.

Ans. type 'icmp' (without quotes) in the filter box, and then click 'apply'

17. Now ping to a name (like www.linux-trailling.be) and try to sniff the DNS query and response. Which DNS server was used ? Was it a tcp or udp query and response ?

Ans.
```
First start the sniffer.
Enter 'dns' in the filter box and click apply.
root@ubuntu910: -# ping www.linux-training.be
PING www.linux-training.be (88.151.243.8) 56(84) bytes of data.
64 bytes from fosfor.openminds.be (88.151.243.8): icmp_seq=1 ttl=58 time=14.9 ms
64 bytes from fosfor.openminds.be (88.151.243.8): icmp seq=2 ttl=58 time=16.0 ms
“C
---www.linux-training.be ping statistics---
2 packets transmitted, 2 received, 0% packet loss, time 1002ms
rtt min/avg/max/mdev = 14.984/15.539/16.095/0.569 ms
```

The wireshark screen should look something like this.

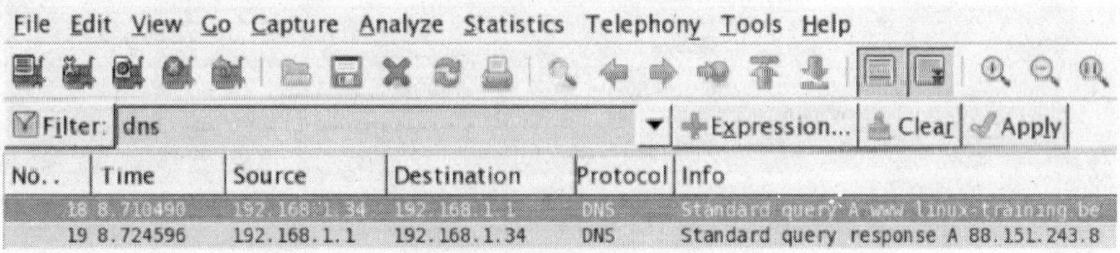

The details in wireshark will say the DNS query was inside a udp packet.

EXERCISE

1. Change **/etc/inittab** so that only two mingetty's are respawned. Kill the other **mingetty's** and verify that they don't come back.
2. Use the Red Hat Enterprise Linux virtual machine. Go to runlevel 5, display the current and previous runlevel, then go back to runlevel 3.

3. Is the sysinit script on your computers setting or changing the PATH environment variable ?
4. List all init.d scripts that are started in runlevel 2.
5. Write a script that acts like a daemon script in **/etc/init.d/**. It should have a case statement to act on start/stop/restart and status. Test the script!
6. Use **chkconfig** to setup your script to start in runlevels 3,4 and 5, and to stop in any other runlevel.
7. Schedule two jobs with **at**, display the **at queue** and remove a job.
8. As normal user, use **crontab -e** to schedule a script to run every four minutes.
9. As root, display the **crontab** file of your normal user.
10. As the normal user again, remove your **crontab** file.
11. Take a look at the **cron** files and directories in **/etc** and understand them. What is the **run-parts** command doing ?

❐❐❐

Solved Model Questions

Q.1. How large a swap partition should I create?

Ans. Although it can be smaller, for best res ults the partition size should be at least equal to the amount of memory installed in the system—preferably twice the amount of physical memory. In other words, if you have 64MB of RAM installed use double that amount for your swap space partition (64MB * 2 = 128MB).

If you need more than 128MB of swap space, but you are using an older distribution that does *not support a swap file larger than 128MB, there are ways around the 128MB swap file* limitation: 1) Linux allows more than one swap partition on a hard drive, and 2) Linux also allows swap files to be put in data partitions (i.e., in the same partitions as your programs and data files).

If you create a swap partition and then later decide that it is not large enough for your needs, you can delete the original one and replace it with a larger swap partition if you have the contiguous space to do so—in other words, if there is no other partition immediately following the original swap partition. If there is insufficient space to create the larger swap partition, you can still create a second partition elsewhere on that drive. They do not have to be back-to-back. (Or you can use a disk partitioning program, such as Partition Magic, to reallocate partition space on the fly between existing partitions, without losing data.) Linux supports up to eight swap partitions.

Alternatively, if you have multiple physical disk drives, splitting your swap space between the disk drives may improve system performance, because both drives can be used concurrently for swapping. (This performance increase is more apparent on a system with SCSI drives, because SCSI devices run in parallel—at the same time—while standard IDE devices run in serial—they alternate. Even so, on a system with multiple IDE drives, it can be faster to switch back and forth between physical drives (an electronic operation) than to s huttle the read/write heads of one drive all over the disk, dividing their time between partitions on the same drive.). Also, drives added to the system later may be faster than the original drive that came with the system. Using these newer drives will give you faster disk read/writes when swapping.

Dividing the swap s pace across multiple drives may also increase system reliability. Like any device, hard drives have a finite life expectancy, so spreading the workload between multiple drives means that the primary drive doesn't have to work as hard, which in turn may (no guarantees, though) extend its useful life. (For instructions on how to do this, refer to the question, How can I create multiple a swap partitions on one or more drives?, below.)

Q.2. How can I speed up performance by using multiple swap partitions?

Ans. If your system (especially a server) has multiple drives set up as a RAID 0 array (data striping), it would automatically spread the data across all the drives in parallel, greatly increasing the speed of disk writes versus a non-RAID setup, where the swap partition is on only one drive. Of course, not everyone can afford a RAID setup; fortunately, there is an alternative that offers many of the same performance benefits in a multidisk environment, at least as far as swap space performance is concerned.

As it happens, there is a way to "parallelize" swap file read/writes through the use of the priority setting in the /etc/fstab file. (Note: This works best with multiple SCSI drives and/or controllers. Two IDE drives on the same controller will see little or no performance boost from this technique, but multiple IDE drives on multiple controllers may benefit somewhat.)

Use an appropriate tool (such as kfstab [available from kfstab.purespace.de/kfstab], which knows the exact layout of the file that fstab requires) to open the file /etc/fstab. This file lists the partitions used by Linux, including swap space. (Note: It is possible to use a simple text editor such as vi, but if you do not get the columns lined up just right, or if you use tabs instead of spaces or vice versa, you can corrupt your system; so if you elect to go this route, rather than using a tool specifically designed for editing fstab, such as kfstab, be sure to mak e a backup copy of fstab first, so that you can recover if your system crashes.)

When you open fstab, if you have multiple SCSI drives with a swap partition on each, you might see something like this (without the headings):

```
<partition  <mount  <partition  <mount  <backup
name>  point>  type>  options>  dump>  <pass>
/dev/sda2  none  swap  sw  0  0
/dev/sdb2  none  swap  sw  0  0
/dev/sdc2  none  swap  sw  0  0
/dev/sdd2  none  swap  sw  0  0
```

Under normal circumstances, Linux would use the swap partition /dev/sda2 first, then / **dev/sdb2, and so on, one at a time, until it had enough space to meet its current swapping** needs.

On the other hand, if you change the settings to:

```
/dev/sda2  none  swap  sw,pri=3  0  0
/dev/sdb2  none  swap  sw,pri=3  0  0
/dev/sdc2  none  swap  sw,pri=3  0  0
/dev/sdd2  none  swap  sw,pri=1  0  0
```

you will be assigning the first three partitions the same priority level—a higher one than the fourth partition. (Swap partitions are used in order from highest priority to lowest—where a priority level of 32767 is the highest and 0 is the lowest.) This has the effect of forcing Linux to write to the first three partitions in parallel, greatly increasing the read/write throughput of the swap space. The fourth partition would be used only if the first three fill up.

In addition to parallelizing the swapping, this technique also allows you to prioritize access to the fastest drives first. Therefore, you can set up an emergency swap partition on an old, slow drive, using low priority, so that it will be used only when all high-speed swap space is exhausted.

Q.3. How can I create multiple swap partitions on one or more drives?

Ans. To create multiple swap partitions:

1. First, use the: su – command to enter superuser mode. (If you don't do this you will be unable to use the fdisk command.)
2. Enter the command: fdisk /dev/hda if your bootable hard drive is IDE, or us e: fdisk /dev/sda if your bootable hard drive is SCSI. (If you are not sure which type of disk drives your system uses and you guess wrong, the worst that can happen is that Linux will issue an error message saying "Unable to open xda." You can't hurt anything.) This is a text mode program, so it is not pretty. Where it says, "The number of cylinders for this disk *is set to xxxx" note the number displayed for xxxx (for example, 1024).*
3. At the prompt "Command (m for help)."

   (*a*) **Type:** p (and press Enter) to see the current partitions. Check the number shown under the End column for the last cylinder listed. If it is less than the number of cylinders from Step 2, above, there is room for another partition. (It is only necessary to create another partition on the same drive if your Linux distribution doesn't allow swap partitions larger than 128MB and you want additional swap partitions.) The difference between the two numbers determines how much space is available for use.

   (*b*) (If there is already an existing swap partition, but it is not large enough, use the: d (delete) command to remove it; then create a new, larger one.)

   (*c*) Enter: n to create a new partition.

   (*d*) If you are creating a Primary partition:

1. Enter: p to create the partition. When prompted for the partition number, use the next higher number than the highest existing partition number. (In other words, if the last device description on the left shows: /dev/hda1, use 2 as the new partition number.)
2. Type in the number suggested for the first cylinder (which should be one higher than the ending cylinder for the preceding partition, or press Enter to take the default.
3. Enter the partition size in megabytes, such as: +128M. (Be sure to include both the + and the M or you will get a "Value out of range" error.)
4. Enter: p again to verify that the new partition was created. It should show an Id type of 83 and a System type of Linux.
5. Enter: t to change the partition type, then enter the partition number you used in Step 1), above. Finally, when prompted for the "Hex code," enter the swap partition type code: 82. If everything worked correctly, you should see a message like "Changed system type of partition [x] to 82 (Linux swap)." (where [x] is the partition number from Step 1).

(e) If you are creating an Extended partition:

1. Enter: e to create the partition. When prompted for the partition number, use the next higher number than the highest existing partition number. (In other words, if the last Device description on the left shows: /dev/hda1, use 2 as the new partition number.)
2. Type in the number suggested for the first cylinder (which should be one higher than the ending cylinder for the preceding partition, or press Enter to take the default.
3. Press Enter to reserve all the available space.
4. Enter: p again to verify that the new Extended partition was created. It should show an Id type of 5 and a System type of Extended.
5. Enter: t to change the partition type, then enter the partition number you used in Step 1), above. Then, when prompted for the "Hex code", enter the Linux Extended partition type code: 85. If everything worked correctly, you should see a message like "Changed system type of partition [x] to 85 (Linux extended)." (where [x] is the partition number from Step 2). Note: This step is not strictly necessary. You can create Linux swap partitions in a "regular" extended partition, but changing the partition type to Linux Extended takes only a few seconds and it clarifies the partition structure for future fdisk users on this system.
6. Enter: n to create a new partition. When prompted for the first cylinder number, press Enter. Enter the desired swap partition size in megabytes, such as: +128M. Enter: p again to verify that a type 83 partition was created with the same starting block as the Linux extended partition you created in Step 1).
7. Enter: t to change the partition type, then enter the number for the Linux extended partition you just created. When prompted for the "Hex code," enter the swap partition type code: 82. If everything worked correctly, you should see the message "Changed system type of partition [x] to 82 (Linux swap).
8. If you did not use up all of the available partition space in Step 6), create one or more additional partitions to hold data, but this time do not change the partition default type of 83.

(f) To write your changes to the hard disk and ex it the program, type: w and press Enter.

1. When you are returned to the Hard Disk Selection panel, select No further hard disk *changes and press Enter.*
2. Repeat steps 2-4 for each additional hard drive that you wish to add a swap partition to, changing the fdisk command from: fdisk /dev/hda to: fdisk /dev/

hdb; or from:

fdisk /dev/sda to: fdisk /dev/sdb, depending on the type of hard drives you are using. When you are done, enter: q to exit fdisk. (For more on disk drive naming conventions in Linux, see question Wait a minute! Are you saying that there are no drive *letters in Linux?)*

Note: To see the complete list of partition types available for your Linux distribution, at the "*Command (m for help)" prompt, enter: l (for "list partition types").*

For performance reasons, if you have more than one hard disk drive your first choice should be to create multiple swap partitions, rather than multiple partitions on the same drive or one swap partition and a swap file in a data partition. However, if you have a shortage of available Primary partitions (or there is no more unallocated disk space from which to create new partitions), you can create swap files in existing data partitions. To do this, see the question *How do I create a swap file in an existing Linux data partition?, below:*

Q.4. How do I create a swap file in an existing Linux data partition?

Ans. If for some reason you can't, or do not wish to, create Linux swap partitions—or if you have no more available disk space to us e for a new swap partition—it is possible to create swap files within partitions that currently hold programs and/or data files. However, your first choice should be for one or more Linux swap partitions.

Important Note: There are two disadvantages to locating swap files outside of dedicated swap partitions: 1) The performance of swap files that are located in data partitions is slower than that of swap files in swap partitions (due to file system overhead and noncontiguous data block s), and 2) It is also possible that if the swap file is ever damaged it could corrupt the filesystem for that partition, resulting in lost data. For these reasons it is strongly recommended that you use swap partitions whenever possible, leaving swap files in data partitions as a last resort. Swap files might be appropriate if you temporarily need some extra swap space.

You can create as many as eight swap files, if necessary. Linux will apportion the swapped data between these files as appropriate.

To create a s wap file in a data partition, use a command similar to the following:

```
dd if=/dev/zero of=<swap path> bs=<size> count=<size>
```

For example:

```
dd if=/dev/zero of=/Swapdir/Swapfile bs=1024 count=65536
```

(**Note:** If /Swapdir does not yet exist, it must be created before running the dd command. For example: mkdir /Swapdir or mkdir /home/Swapdir. In order to create the directory, you must be running as the root operator.)

The dd command (sometimes called "data duplicator," although the original meaning of the dd acronym is lost in the mists of time) is used to create the swap file. The if= parameter identifies the input file source for creating the swap file (device "zero"), of= is where you specify the path to the output (swap) file you want to create, bs= is the block size to use and **count= is where you define how large to make the swap file, in increments of the block size.** In the preceding example, we are creating a 64MB swap file (65,536 * 1024 [1K] equals 65,536K, or 64MB), called Swapfile (use whatever name you prefer), in a directory called **Swapdir.**

Next, to initialize (or make ready for use) the swap file, use the following command:

```
mkswap /Swapdir/Swapfile 65536
```

(Use the same path as in the dd command and the file size you specified in the count= parameter. If you get a "No such file or directory" error, verify that you typed the upper/ lower case correctly and did not omit the leading /.)

After initializing the file, "synchronize" it to ensure that it is properly written to disk:

```
sync
```

And, finally, to tell Linux to start using this new swap file:

```
swapon /Swapdir/Swapfile
```

This will activate the swap file temporarily (until the next time you log out or reboot). To make it permanently active, use an appropriate tool (such as kfstab [see the earlier question, How *can I speed up performance by using multiple swap partitions?, for details], which knows the* exact layout of the file that fstab requires) to open the file /etc/fstab.

When you open fstab, you might see something like (without the headings):

```
<partition <mount <partition <mount <backup
name> point> type> options> dump> <pass>
/dev/hda1 / ext2 defaults 0 1
/dev/hda2 /usr ext2 defaults 0 1
/dev/hda3 none swap sw 0 0
```

(There may be other disk partitions listed if you have other swap and/or data partitions.) To activate your new swap file, add a line similar to the following immediately after the last partition listing. Then save and close the /etc/fstab file.

```
/Swapdir/Swapfile none swap sw 0 0
```

Should you later decide that you no longer need the swap file, it is easy enough to remove. First, use the swapoff command to disable the use of the file:

```
swapoff /Swapdir/Swapfile
```

Then, simply delete the file:

```
rm /Swapdir/Swapfile
```

Q.5. I already have a large swap file in my Windows partition. Is there a way for Linux to use that swap space instead of creating another file?

Ans. Yes, although Linux is rather meticulous about what format the data takes in its swap space. Thus , there is more to it than simply pointing Linux to the path of the Windows swap file.

I have found several descriptions of how to accomplish this (and all of them reportedly work) but I have not had the need to try any myself. They all appear to be terribly complicated, so unless you consider yourself well versed in Linux, it might be best to avoid this topic entirely until someone develops a utility to simplify the process. If you would like to look over one such procedure, to mak e up your own mind whether or not to try it, read the document at *http://www.linuxdoc.org/HOWTO/mini/Swap-Space-1.html. It seems rather thorough. (Use the* left and right arrows to move back and forth through the document. Use the up arrow to go to the document table of contents .)

(If you already have a Linux s wap partition before you decide to try this technique to use your Windows swap file instead, you should later delete it using fdisk, to free up the disk space used by that partition. For directions on how to use fdisk to check your partition tables, and to delete a partition, see the question How large a swap partition should I create?, above.)

Another possible way to share swap space is to have a swap partition shared by both operating systems . This is not a "no-brainer" either, but if you are interested do an Internet search for "Linux swap space" and you should find several such documents.

Q.6. If I install Linux on an IBM® ThinkPad® with a TrackPoint® II or III pointer, what kind of mouse should I select?

Ans. The installation program should correctly recognize the Track Point II/III as a PS/2-type mouse.

Most such ThinkPad notebooks have two "mouse" buttons, but some newer systems have a third button. When installing on a ThinkPad with two buttons, select the Emulate 3 Buttons option on the Mouse Configuration screen. If your Think Pad has three buttons do not select this option, or it will cause problems. To use the "third button" on a two-button ThinkPad (or any other system with a two-button mouse), press both buttons together (called "chording"). (These instructions should apply equally to other brands of notebook computers using IBM's "pointing stick" technology, identified as a little "pencil eraser"—usually red, green, or black—nestled between the G, H and B keys; but to be sure, contact the notebook vendor.)

Migration from Windows : If you are in the process of moving from Windows to Linux, or are setting up a Linux system to coexist with Windows computers and perhaps share data between the two types of systems, these questions may save you some time and effort.

Q.7. How do I get Netscape for Linux to recognize my Netscape for Windows bookmark file?

Ans. For best results, try to use the same version of Netscape on both systems, or as close as possible (e.g., 4.72 and 4.7, but not 3.0) to minimize the chance of inconsistencies in the bookmark file format.

Use the Linux Kfind, or similar, utility to determine where the default bookmark file is kept. It should be in a directory with a name like /home/username/.netscape (substitute your account name for username). If you simply copy the bookmark.htm file from your Windows PC to a floppy and then to the appropriate Linux directory, when you restart Netscape and open the book marks you will see the default bookmark s, not your customized list. The problem is in the file name. The Linux version is called book marks.html (note the two extra letters). Therefore, you must rename the bookmark.htm file to bookmarks.html before Netscape for Linux will recognize it (reverse the process to copy an updated file back to your Windows system).

First, copy the bookmark.htm file from the Windows system. If you use a floppy, Zip disk or other removable medium, you can rename the file on the removable disk before copying to the Linux system. (The file should be stored in a directory called something like \Program **Files\Netscape\Communicator\Users\yourusername, but you can use the Windows Find or** *Search function—depending on the Windows product—from the Start menu to locate it.) Once* the file is copied to the removable disk, use the Windows file manager to rename it **bookmarks.html.**

On the Linux computer, KDE graphical interface users can copy the file to the **/home/ username/.netscape directory using the K File Manager (just click on the icon on the** tas kbar, or Panel, that looks like a house in front of a manila folder). Gnome users can use the Gnome File Manager (click on the Gnome icon—the vaguely G-shaped footprint—on the Panel, select Programs, then File Manager). If you prefer to work from a command line interface, use the following command:

```
cp /dev/fd0/bookmarks.html /home/yourusername/.netscape
```

(Note the use of forward slashes and that /dev/fd0 is used in place of a:. Linux and UNIX® do not use drive letters, unlike DOS, Windows or OS/2.)

Alternatively, you could combine the renaming and copying steps into one, using the following command:

```
mv /dev/fd0/bookmark.htm /home/yourusername/.netscape/bookmarks.html
```

(**Note:** The diskette drive may already be "mounted" as /dev/floppy or /mnt/floppy in your Linux distribution. If so, use that mount location instead of /dev/fd0.)

If you transfer the file directly to the Linux system (via network connection, for example) or prefer to rename the file once it is on the Linux system go to the /home/username/.netscape directory (cd /home/username/.netscape) and use a file manager or the move (mv) command to change the name: mv bookmark.htm bookmarks.html. (If you use the mv command without a target location, it simply renames the files.)

Note: In addition to the bookmarks file, you might also want to copy the following files from the same Netscape directory on your Windows system, to your Linux system: liprefs.js, prefs.js and proxy.cfg. They contain the user preferences information (font selections, proxies, servers and so on). In addition, if you did not have "cook ies" disabled on your Windows system, copy cookies.txt as well (some Web sites store information about your user account in this file). This will save you from having to manually reenter your preferences on your Linux system. (You will probably have to overwrite the default files with the same names on your Linux system.)

If Netscape is still running, close and res tart it so it will read the new bookmarks file (as well as the other files you copied). If it still doesn't work, verify that you put it in the correct directory and have spelled the file name accurately—it must be in all lower-case.

Q.8. Are there any Linux programs that will read and write Microsoft Office files?

Ans. Yes, there are. For a list of several such programs and how compatible they are, see the white paper entitled What Good is a Linux Client?, available from the same sources as this paper.

Configuration & Setup : Here are some tips for tailoring Linux or configuring devices.

Q.9. How do I get my "winmodem" to work with Linux?

Ans. Winmodem is actually a trademark of 3Com Corporation, but is often used generically to describe a class of low-cost software-based modems, designed to be inexpensive by replacing some hardware functions with software. This software typically incorporates low-level Windows functions that can't be duplicated in Linux. Most such modems won't work with Linux, however, some can be made to be compatible. These sometimes are referred to as "Linmodems". Go to http://www.o2.net/~gromitkc/winmodem.html for more information on this topic. In most cases, the easiest "fix" is simply to disable the software-based modem and install a traditional modem adapter, or attach en external modem to a serial or USB port. Many 56K modems are quite inexpensive these days. (For information on how to disable the built-in modem, consult the user's guide for the computer, or contact the computer vendor's technical support organization.)

Q.10. How can I tell how much memory Linux is using?

Ans. From a command s hell, use the "concatenate" command: cat /proc/meminfo for memory usage information. You should see a line starting something like: Mem: 64655360, etc. This is the total memory Linux think s it has available to use. If you have more than 64MB of RAM installed and see a number c lose to the one above, Linux is not using the remainder. To see how much memory is currently unused, use the free command.

Q.11. If Linux is not using all the memory I have installed, how do I make it use the rest?

Ans. To force Linux to recognize and use the full amount of memory you have installed, open a command shell and login temporarily as the "root" operator, with the su command. (For instructions on how to do this, see the Root/User Accounts, Groups and Permissions section, below.) When prompted, type in the root password. At the command prompt, use an editor to open the file /etc/lilo.conf, such as: pico /etc/lilo.conf. The first line of the file should start: boot=, followed by other lines for map=, install= and so on. Position the cursor in front of the "b" in boot=, then insert the following: append="mem=128M" and press the Enter key to move the boot= line down. Finally, close the editor and exit the file (Ctrl-X, in the case of the Pico editor, Q to quit various other editors). Save the changes, if prompted. From the command prompt, type: /sbin/lilo to restart LILO, then type: exit to logout as the root operator and then exit again to c lose the shell session. The next time you reboot Linux should be using all the memory installed. (Use the cat /proc/meminfo command to verify.)

Warning: It is possible to render Linux unbootable by spec ifying a memory size larger than that physically installed and available. If your computer reserves some memory, rendering it unavailable to Linux, you must specify the lesser amount. For example, during power up (POST), if your memory count shows something like 130496 (kilobytes), rather than the full **131072 KB for 128MB of RAM (128 * 1024), this means that some of the memory is reserved** by the hardware. In this case, do not use mem=128M in the example above. Instead, use the exact amount of 130496K (append="mem=130496K"), or play it safe and "step down" to 127MB, as in: append="mem=127M". If you somehow manage to leave your system in an unbootable state (i.e., it locks up on boot when it tries to us e the reserved memory), the remedy is to edit the /etc/lilo.conf to correct the memory size.

Q.12. I have a printer attached to the parallel port. What is this port called in Linux?

Ans. When prompted for the name of a parallel port in Linux, in place of LPT1, LPT2, or LPT3 use **/dev/lp0, /dev/lp1, or /dev/lp2, respectively.**

Q.13. I have a device attached to a serial port. What is this port called in Linux?

Ans. When prompted for the name of a serial port in Linux, in place of COM1 through COM8 substitute /dev/ttyS0 through /dev/ttyS7, respectively.

Q.14. What is my floppy drive called in Linux?

Ans. If prompted for the path of a diskette drive in Linux, in place of "a:" substitute /dev/fd0, **/dev/floppy, or /mnt/floppy, depending on the Linux distribution you are using. Unlike the** DOS/Windows world, UNIX/Linux doesn't use drive letters. Devices are assigned "mount points" (or mounting locations). To determine which mount location is assigned to the floppy drive on your system, type: mount at a command prompt. You should see something like:

```
/dev/hda1 on /type ext2 (rw)
none on /proc type proc (rw)
none on /dev/pts type devpts (rw,gid=5,mode=620)
/dev/fd0 on /mnt/floppy type vfat (rw,nosuid,nodev,user=username)
```

Look for the part that starts with /dev/fd0 on.... This tells you what "soft" mount location to use, ins tead of /dev/fd0: "/mnt/floppy", in this particular c ase.

Q.15. Wait a minute! Are you saying that there are no drive letters in Linux?

Ans. That is correct. Hard drives, floppy drives, CD-ROM drives , Iomega Zip and Jaz drives, and other devices that get drive letters in DOS/Windows are instead assigned mount points by Linux. Depending on the device involved, here are some of the designations used by Linux, following their typical Windows drive letters:

Floppy drives

A: drive (in Windows) = /dev/fd0 (or /dev/floppy or /mnt/floppy in Linux, as explained above)

B: drive = /dev/fd1 (see note, below)

IDE/EIDE disk drives

C: drive (boot drive; first/primary drive on the first disk controller) = /dev/hda

D: drive (second/slave drive on the first disk controller) = /dev/hdb

E: drive (first/primary physical drive on the second controller) = /dev/hdc

F: drive (second/slave physical drive on the second controller) = /dev/hdd

(CD-ROM drives will often be attached as the primary drive on the second controller, or **/dev/hdc. IDE/ATA Zip drives will generally take the next available drive letter after the hard** drives and/or CD-ROM drives.)

SCSI disk drives

C: drive (boot drive; lowest-numbered drive on the primary controller) = /dev/sda

D: drive (second lowest-numbered drive on the primary controller) = /dev/sdb

E: drive (third lowest-numbered drive on the primary controller) = /dev/sdc

(And so on. There can be many SCSI drives, on one or more controllers. The mount point designations just keep incrementing up through /dev/sdz. If there are more than 26 physical drives, they will be incrementally labeled from /dev/sdaa up through /dev/sdz z.)

Note: Unlike the primary floppy drive (/dev/fd0) Linux does n't automatically map a second diskette drive (/dev/fd1) to something like /dev/floppy or /mnt/floppy. However, if desired you can set up your own alias, such as /mnt/floppy2. The mapping entries that are used for this are located in the /etc directory, in a file called fstab. To add an alias for fd1, use an editor to open /etc/fstab and look for the entry for fd0, which s hould look something like:

```
/dev/fd0 /mnt/floppy auto noauto,owner 0 0
```

This line maps /dev/fd0 to the directory /mnt/floppy. To create an alias for fd1, create a subdirectory in the /mnt directory called floppy2. Then add a new entry (using whatever alias you wish) to the fstab file in order to map the second floppy drive to a mounting point, such as:

```
/dev/fd1 /mnt/floppy2 auto noauto,owner 0 0
```

Save the file and close the editor. Now, when you us e the Mount command it s hould include something like:

```
/dev/fd1 on /mnt/floppy2 type vfat (rw,nosuid,nodev,user=username)
```

Q.16. What about disk drive partitions? Don't they have drive letters either?

Ans. Nope. Linux just tacks numbers onto the end of the drive identifier. For example, if the first IDE hard drive had three primary partitions, they would be named/numbered, /dev/hda1, **/dev/hda2 and /dev/hda3. If a second hard drive had two primary partitions , they would be** called dev/hdb1 and /dev/hdb2. Likewise, a SCSI hard drive with three primary partitions would be labeled /dev/sda1, /dev/sda2 and /dev/sda3.

That is just for primary partitions. If you also have extended partitions, they begin numbering at 5 and continue from there. If, for instance, you had one IDE drive with two primary and two *extended partitions, the partitions would be called: /dev/hda1, /dev/hda2, /dev/hda5 and* **/dev/hda6.**

Q.17. Wow, that's confusing! Isn't there a simpler way to access drives?

Ans. Yes. Most Linux distributions provide desktop icons for some devices , such as floppy and CD-ROM drives that can be clicked on for direct access. If you install software for accessing an Iomega Zip drive, it should also provide an icon to use. And, of course, the graphical file managers, such as the K File Manager and Gnome File Manager allow you to access files on various drives without worrying about drive letters. (When you open KFM, look for an icon called mnt/. When you single-click on it, KFM will display all the devices that are ready for use. Click on the icon called cdrom/ to display the contents of the currently inserted CD, or click on the one called floppy/ to see what is on the inserted diskette.)

Q.18. Is there an easy way to see what devices, such as drives, have been mounted?

Ans. Yes. Red Hat users running KDE can click on the KDE menu and select Red Hat, then *System and finally Disk Management. This will bring up a panel that shows the devices* mounted, and provides Mount and/or Unmount buttons (so that you can take a device offline, then remount it). The Mount/Unmount buttons will toggle back and forth between the two options being used. There are also buttons provided to let you format the drives.

Gnome users (of any distribution) can select File manager from the Gnome icon on the Panel (the Linux "taskbar") and then click on the /mnt icon to view the various devices.

I have not seen a KDE equivalent for the Red Hat Disk Management tool for other distributions. As an alternative, the Mount command can be used, as described earlier, in the question What is my floppy drive called in Linux?. Also, for a less cluttered view of disk drives mounted, with total and available space, try the disk filespace command (use: df –h to display measurements in "human-readable" form—gigabytes, megabytes, etc.). It won't show non-disk devices, however. (Read the info manual pages at: info df, for more information.)

Q.19. Is there any way to change the default language used by the operating system and applications?

Ans. The short answer is "Yes." Of course, as with most things Linux, how you do it depends on the distribution you are using, the user interface (desktop environment) you are using, your applications and perhaps other factors. The default language is set during installation, however there are ways to change this once the system is up and running.

If, like many users, you are running the KDE environment, the way to change the language is very simple. From the KDE menu:

1. Select Settings, then Desktop, then Language.

2. Specify your first choice from nearly three dozen languages (including Catalan, Slovak and Macedonian). You are allowed to choose up to three languages. (When you start a program, KDE will try to find a translation for the menus, dialog boxes and messages in the first language. If it fails, it will try the second choice and, if necessary, the third. If all three fail (or if you do not select three languages), it will use whatever was the original default language specified during installation.)
3. Press OK to save your changes and exit. (Apply will save your changes but not exit. The *Default button will return all three language settings to the original "Default language"* choice.)
4. Log out and then log in again to see your changes.

Note: This procedure only affects certain system utilities, help files, error messages and the like. Your applications may or may not be affected, depending on what support for other languages are programmed into them. Therefore, you may end up with a mix of two or more languages depending on the various programs you are running at any one time. You may be able to set up language choices within specific applications as well.

If, after logging out/in you discover to your horror that you somehow selected the wrong language and you can't read most icons, menus and other system functions, try to follow the instructions above. If you c an manage to return to the language settings panel, but can't read the language names, the little flag icons should allow you to find the language you want, or you can press the Default button at the bottom—the second one from the left, if you can't read it. If you are unable to find your way back to the language settings you may have to reinstall *Linux to specify the default language all over again.*

Gnome users can use the same KDE tool, if they wish. The only extra step is to go to the Gnome menu, select the KDE menus option, then proceed as directed, above.

If you use another desktop environment besides KDE or Gnome, refer to the manual or tutorial that came with the environment for information on how to change the language.

General Usage: How to perform various task s using Linux.

Q.20. How do I format a floppy (diskette)?

Ans. If you are using the KDE user interface, the easiest way is to use the KDE Floppy Formatter utility supplied. Simply click on the KDE icon on the Panel, select Utilities, then KFloppy. The File System option gives you the choice of the Dos (FAT) file system (for DOS/Windows or OS/2) or the Linux native ext2fs file system.

If you are using Gnome, go to the Gnome icon (the fancy G-shaped footprint) on the Panel and select Utilities, then GFFloppy.

From the command shell, to format a diskette for the native Linux file system, use the mke2fs command. You can use the mtools utility to access DOS/Windows-formatted diskettes. Use the commands: mformat, mcopy and mdel without parameters to see a list of command options. (Using mtools, to change from one directory on the floppy to another, use the following command: mcd a:subdir, where subdir is the name of the directory you wish to change to.)

Q.21. Is there online documentation for most programs?

Ans. Yes. The official help tool for Linux is called man (short for manual). To read the manual for various programs (including many commands), type: man <something> (where *<something> is the name of the program you need help with) at a shell prompt. For example,* to read the online manual for the man program itself, use: man man. You can also use a viewer called less to view the manual: man man | less (where | is the piping symbol—**Shift-Backslash—located above the Enter key on a 101-key keyboard). Using less, press** the Spacebar to move ahead a page at a time; to back up, use the B key. When you are done, press Q to quit.

Many newer programs have shifted to the info tool rather than using man. Like man, you can view info pages using the command: info <something>. Some programs now supply HTML or XML pages, kept in the /usr/doc/<programname> directory, in place of either man or info pages. These documents may be viewed from a Web browser or other HTML/XML-capable program.

Another option is to redirect the manual to a tex t file for browsing with an editor. For example: man mv | col –b > move.txt (to redirect the text for the mv command) will strip out the screen formatting codes that do not translate into print and then copy the output to a file called **move.txt (col is the name of the formatting utility). Once it is in a text file, it can be read with** Emacs, Pic o, Vi, or any other text editor. (Note: Substituting >> for > in the example above would append the text to the end of an existing file, whereas using > a second time would *overwrite man.txt with new data.)*

Likewise, to print a manual you will need to pipe (send) the document to a formatting program that will strip out the screen formatting codes before printing. To do this for info pages, type: info mv | col –b | lpr. (lpr is the program that sends the document to the print spooler.)

Another "quick and dirty" source of helpful information is the what is command. What is provides a brief synopsis of many commands. For example, try: what is man. You will be shown several "definitions" of what the man command does. The same goes for: whatis whereis, to find out what the whereis command does. If there is no information available for a particular command you will see a "nothing appropriate" message instead. Note: If the ***what is command doesn't work, it may mean that the whatis database has not yet been built.***

To create it, first use the: su – command to log in as root, then enter: /usr/sbin/makewhatis. This may take several minutes. When the process finishes, try the ***what is command again.***

Q.22. How do I kill a program that has locked up?

Ans. If you are not the owner of the program (the user account who installed it), you must log in as root. (For instructions on how to do this, see the Root/User Accounts, Groups and *Permissions section, below.) Then, from a command line, you can type: ps (process status) to* see what programs are running. To the right of the process name will be a Process ID (PID) number, such as 506. This is what you will use to kill the program. From the command prompt, type: kill pid (for example, kill 506). This tells the program to either shape up (respond normally) or ship out (close itself). This will usually solve the problem, but occasionally stronger measures are required. Preceding the PID with –9 (such as: kill –9 506) will kill the program. (This is sometimes called "signal 9," not to be confused with "Plan 9 from Outer Space", a cheesy B-movie.)

An alternative for killing a background process, or daemon is to use the –HUP (hangup) parameter, as in: kill –HUP 506. It essentially signals the program to clos e itself, then restart and reread its configuration file. Again, if the kill command doesn't work the first time, you may have to resort to the –9 parameter rather than –HUP.

Q.23. Is there a way to increase the priority of a program to make it run faster?

Ans. Yes. If you have a program that normally takes quite a while to run and you really, really need it to complete faster, you can boost its job priority level so that it uses more process or cycles. Of course, this means that all other currently executing programs will run slower. But if this is what you want, the nice command will do the job for you.

Linux supports priority levels from 19 (lowest) to –20 (highest), with the default being 0. To change the priority to a negative number (increasing the priority), you must first log in as root.

Then, to start a program called Myprog with a high priority-level of –10, use the command: nice –n –10 myprog. If you wish to increase the priority of an already running program, use the renice command, identifying the program by its PID: renice –10 506 (the –n argument is not used).

Conversely, if you have a long-running program that is a low-priority job and it is slowing the entire system down while the program is running, you can lower its priority using the same techniques by providing positive numbers—such as 8—rather than negative numbers. By default, nice will reduce the priority by 10. In other words, the command: nice myprog is the same as: nice –n 10 myprog.

It may take some experimenting for you to find the right priority levels to assign to specific programs. Too high a priority may bring the rest of the system to its knees; too low and the job may take forever to complete. As a rule of thumb, unless you have a reason to be changing *priorities, leave them alone. The default setting is best most of the time.*

Q.24. Is there a spelling checker that I can use outside of a word processor?

Ans. Yes. Linux includes a program called ispell, which will read any text file, highlight questionable words, offer alternatives and let you add the words to the ispell dictionary if desired. It runs from the command line, with a text-mode (i.e., non-graphical) interface, but is fairly easy to use. There are a number of parameters that you can use (type: ispell —? for a list of them or info ispell for more detailed explanations). But the simplest usage to check a file called memo.txt would be: ispell memo.txt. It will even check a series of files in order, such as: ispell memo.txt mydata.txt or even ispell *.txt (to check all files in the current directory with a .txt ex tension).

When the program runs it will pres ent a "questionable" word within the context of the line of tex t containing the word and offer suggested replacements in a numbered list. To select one of the suggestions from the list to replace the original word, simply type in the number of the word you prefer. There are also several menu options displayed along the bottom of the program. To Replace the word with something you type in, type R (or r, ispell options are not case-sensitive) and then whatever word you want to replace the original with. (You also can type in the number of a replacement word from the list. This is the same as just typing in the number directly, but is there in case you forget you can do it that way. Use A to Accept the existing word as is (i.e., ignore it) for the res t of the spell-checking session. (Using the **spacebar tells ispell to accept/ignore the word this time only.) I lets you Insert (or add) the** existing word into the ispell dictionary. Entering an L lets you Look up other words in the dictionary. (Perhaps you would like to find another word not on the suggestion list.) If the word is correct but capitalized inappropriately, choosing U will Uncapitalize it. Either Q or X will end the program. If you have specified a list of files to spell-check, X will stop checking the current file, save it and begin on the next one. This is essentially "save and exit." On the other hand, Q will ask whether you really want to Quit immediately without saving your changes. (Type Y or **N, depending on your preference.)**

Be sure to check out the associated online help for that program by entering ?. In addition to explaining the above options, there are others listed that you might not k now about or remember, including Redraw screen (^L).

> ***Warning: ispell is not discriminating as to what files it checks. It is entirely possible to spell-*** check a word processing document or a binary program file. Saving changes to a document created by a word processor may (or may not) corrupt it, but the word processor probably has its own built-in spelling-checker anyway, so why bother with ispell? On the other hand, you *will corrupt a program if you change anything and then s ave the changes with the X option. If* you accidentally open a binary file (easily identifiable by the "gibberish" displayed that ispell is trying to correct), be sure to Quit immediately with the Q option!

GUI/Desktop : The following questions involve window manager/desktop and other graphical user interface questions.

Q.25. My copy of Linux came with KDE and Gnome. Why do I need both?

Ans. Technically speaking, you probably do not need both. However, both desktop environments are popular, so this gives you the option to decide which you like better. Also, some programs will work only with one or the other, allowing you to use both types of programs, switching between the two desktops as needed.

Q.26. I have both KDE and Gnome installed. How do I switch between them?

Ans. From the graphical interface, simply log out. Then at the Log in screen, type your login ID and password and choose which desktop you wish to load (Session Type). This choice will remain your default until you change it to something else.

Q.27. Can I add more desktop themes to use with Gnome and KDE?

Ans. Yes. One source for more Gnome themes is http://gtk.themes.org. To install a new theme in Gnome, after downloading it, start the Gnome Control Center by clicking on the "toolbox" icon on the Panel. Then click on Theme Selector (under Desktop) in the panel on the left. Press the *Ins tall new theme button (in the center of the right-hand panel), and when the dialog box* opens, scroll down to the name of the file you downloaded (called something like **new_theme.tar.gz and most likely in your login directory). Click on the name of the theme you** downloaded and press the OK button. When the theme appears in the Available Themes box you can press the Preview button (if Auto Preview isn't already selected) to see how it looks before applying it. If you think you might like it, press the Try button to conditionally apply it to your desktop. If you approve of the results, press OK to save the changes. If you decide you do not like it after all, use the Revert button to back out the theme and return to your previous theme.

KDE themes can be found at http://kde.themes.org. To install a new KDE theme, start the KDE Control Center from the "home" icon (it looks like a little house in front of a manila folder) on the Panel. Then, click on Theme Manager (under Desktop) in the panel on the left. Press the Add button (in the upper-right part of the right-hand panel). When the dialog box opens, find the name of the theme file you downloaded. Double-click on the name to install it. Then click on the theme name in the Installer box. The window in the middle shows how it will look. If you like the appearance, press the Apply button at the bottom of the panel to change your desktop theme. If you do not like the appearance, press Cancel to exit the KDE Control Center.

Other sources for themes include many Web sites that offer software downloads. For a list of some of these sites, refer to the What Good is a Linux Client? white paper, available from the same sources as this Q&A paper.

Q.28. How can I add programs to the Panel?

Ans. KDE users can c lick on the KDE icon, select Panel, then Add Application and then choose the desired program from the various menus shown. Once the application has been added to the panel, you can rearrange the order of the icons by right-clicking on an ic on and s electing *Move, then dragging the icon to wherever you want it.*

Gnome users mus t first log in as root, then click on the Gnome menu icon, then Programs and *Settings, then select Menu editor. When the popup menu appears you can create a submenu* or add to an existing menu.

Q.29. What is a virtual desktop?

Ans. When running a number of programs concurrently, the screen sometimes becomes cluttered with several windows open at once, covering each other up. One solution, of course, is to minimize some of the windows until they are needed. However, this entails minimizing and restoring those windows repeatedly, as different windows are accessed. This can be tedious over the course of a day.

An alternative is to use a program that creates "virtual desktops." Each desk top is a clean slate where you can open one or more programs. Rather than minimizing/restoring all those programs as needed, you can simply s huffle between virtual desktops with programs intact in each one. (If this doesn't seem like much of an improvement, consider that with eight open programs, a user might have to minimize and restore each of those eight programs multiple times in turn. With eight virtual desktops, each program could stay open at all times, needing only one click to change to the next, instead of two.) Virtual des ktops c an help reduce desktop clutter.

There are several programs of this type available for Microsoft Windows users to buy or download, but Linux users do not even have to look for such a program. Both KDE and Gnome include virtual desktop support by default and users can specify how many desktops they want to use—or disable them entirely. To switch between desktops, click on the buttons on the Panel (for One, Two, Three, etc.), start one or more programs, then select another virtual desktop, start more programs and so on. To cycle through all the desk tops in turn, use *Ctrl-Tab.*

Q.30. You said I could have eight virtual desktops, but I see only four. Where are the rest?

Ans. The default setting is four, but you can easily change the number in increments of two (i.e., 2, 4, 6 or 8). To do this in KDE, click on the KDE icon on the Panel, then select Panel and *Configure. This will display the KPanel Configuration window. Use the Visible slider to change* the number of virtual desktops. (The Width slider lets you control how big the icons are on the Panel.)

If you are using the Gnome environment, from the Gnome icon on the Panel go to Program and select Setting, then Gnome Control Center. When the Gnome Control Center panel appears, click on Desktop, then Windows Manager and finally Workspaces. Select the desired number of virtual columns or desktops and press OK to close the panel.

Q.31. Is it possible to share a program across multiple virtual desktops?

Ans. Yes. In the upper left-hand corner of a program window there is an icon that looks like a pushpin. Pressing this button will "pin" that application in place, making it appear in all virtual desktops, in the same position onscreen. To remove it from other virtual desktops, pres s the button again and it will remain on only the virtual desktop from which it was "unpinned."

Q.32. How can I change my desktop background color?

Ans. KDE users can click on the KDE icon on the Panel and select Settings, then Desktop and finally Background. From the Display Settings dialog box you can pick which desktop (from among the virtual des ktops) to modify, then choose one or two colors and/or wallpaper to use. To test out your choice, press the Apply button. When you are finished, press OK to save your changes, or Default to go back to the original system default colors. (Cancel will not abort the changes once you press Apply; it will only close the panel.)

Q.33. Is there a faster way to copy/cut & paste text than to use the pull-down menu?

Ans. Yes. There is a quick way to copy text with the mouse. If you are using a three-button mouse, first position the cursor where you wish to place the text, then select (highlight) the text as you normally would with the mouse. Finally, click the middle mouse button. (If you have a two-button mouse set up to emulate a three-button mouse, press both the left and right mouse buttons simultaneously to achieve the same effect.) The highlighted text will be copied to the cursor position. This should work in most text editors and command line sessions, as well as in word processors and other graphical programs. Alternatively, you can paste the text with **Shift-Insert (hold the Shift key while pressing the Insert key), rather than using the third** mouse button.

For a keyboard method, you can highlight the text (using the mouse or holding down the Shift key while using the left and right arrow keys to "paint" the text) then use Ctrl-C to copy (or **Ctrl-X to cut) the text, then Shift-Insert (or the third mouse button) to paste; s imilarly to the** Ctrl-C/Ctrl-X/Ctrl-V trio used in Windows.)

Note: Because of all the different user interfaces, editors and applications used in Linux, not all of these techniques will work in every program, so you may need to experiment to see what works where.

Command Shell : The next series of questions are specific to using commands and command shells.

Q.34. If I want to issue a command, how do I open a command prompt?

Ans. In Linux parlance, it is generally referred to as a command shell, or simply a shell. Each Linux distribution includes at least one shell. The most commonly used one today is bash (short for Bourne Again SHell), an enhanced vers ion of the older Bourne shell. To open the default shell for your distribution, KDE users can click the Kvt icon on the Panel (the icon that looks like a computer terminal), or press Ctrl-Alt-F1. This will provide a command line interface (CLI) from which you can run commands as needed.

Gnome users can go to the Gnome icon on the Panel, select System and open either a regular or color Xterm (terminal) session, or command prompt window, by clicking on the icons that look like a computer terminal with a large red or multicolored X in front of it, respectively. Alternatively, you can right-c lick anywhere and open a new terminal by selecting *New, then Terminal.*

If you need only a "quickie" command line, just long enough to execute one command and then close itself, press the Alt-F2 keys. When the pop-up panel appears, type the command and press Enter. Instead of a command, you can enter a data file name (with the path necessary to find it) and Linux will launch the appropriate program, based on the file association. (For example, typing: /home/username/.netscape/bookmarks.html will cause the default browser to launch.) If you use a path to a directory instead of a file, Linux will open the directory using the default file manager, such as KFM.

Q.35. What are the Linux equivalents to DOS/Windows commands?

Ans. There are a number of commands in Linux that are similar or identical to either the name or function of DOS commands. Here are some common DOS commands with their Linux counterparts:

Command Purpose DOS Command Linux Command		
Change to the parent directory cd.. or cd ..	(short for change directory) cd ..	
Change to the root directory cd\ or cd\cd/	(note the forward slash)	
Clear the screen cls		clear
Close a command prompt window exit		exit
Compare the contents of two files fc (file compare) diff (difference)		
Copy a file	copy	cp (copy)
Create a directory mkdir or md (make directory) mkdir		
Delete files	del or erase rm (remove)	
Delete directories rmdir or rd (remove dir) rmdir		
Display file contents, a screen at a time	more	more or less
Display the amount of available RAM	mem	free
Display/change the date date		date
Display/change the time time		date
Echo output to the screen echo		echo
Find a file by filename dir or attrib locate		
Find a file by date or other file attributes	dir (not by date) find	
Find a text string in a file find		grep
Format a floppy format		mke2fs (or mformat, for a DOS/Windows floppy)
List files in directory dir (directory) ls (list) or dir		
Move files	move	mv (move)
Rename a file	ren (short for rename) mv	
Show the current filepath cd		pwd (print work ing directory)

Even though some of the commands are the same, the parameters for these commands may differ between DOS/Windows and Linux. For example, let us look at the differences between changing the date and time in Linux versus Windows.

In Windows, at a command prompt you would type: date 12-31-2000 to change the date to December 31, 2000.

Then, to change the time to 2:00 p.m., you would use: time 14:00:00 (or simply time 14:00).

In Linux, on the other hand, first you would have to be logged in as the Superuser (or use the command: su – to temporarily grant yourself that authority), then you would use the command: date 123114002000. Notice that the 14:00 is embedded in the date string, between the day and the year. This is the way Linux wants it. You can't specify seconds using this command—just hours and minutes.

However, if you want to change only the time—but not the date—use the following command instead:

```
date -s 1400
```

or, if you want to specify seconds:

```
date -s 14:00:35
```

(The colons are required when specifying seconds.) In both examples, the –s parameter tells Linux that you wish to set the time.

For more on the su command, see the Root/User Accounts, Groups and Permissions section, below.

Q.36. How do I get help for command parameters?

Ans. There are several options available. First, many common commands will have short "manual" pages built into Linux. You can access them by typing: man, followed by the command name, such as: man mv to see the parameters for the mv (move) command. Some utilities provide more detailed (and sometimes newer) command help using the info tool; for example: info mv. If these tools do not produce results, try following the command with –help or –h, as the only parameter (as in mount -h or mv —help). Some commands, such as: mformat, will display the list of available arguments whenever the command is used without parameters, or with invalid parameters. When in doubt, try them all until you find the one that works.

To learn how to use the info command, type: info info (the tool uses itself to display the tutorial). To learn more about man, use: man man and/or info man. (The tutorials are different, so it is not a bad idea to consult both.) When you are finished reading the help displayed by man or info, press the Q key to quit and return to the command shell.

Q.37. How do I repeat a command?

Ans. If you tend to use the same commands over and over, or recently typed in a rather lengthy set of parameters you wish to use again, there are a number of ways you can repeat a command without having to retype it. Linux stores the last 500 commands in a file (called .bash_history) for quick retrieval. The simples t way to find a recent command is to use the up and down arrow keys to scroll back and forth through the list until you find the one you want. If you wish to change a command you used recently, use the up/down arrow keys until you find the command you want, then use the left/right arrow keys to go to the part you wish to change. Type in anything you need to add and use the Del key to delete unneeded text.

Alternatively, typing: !! (pronounced "bang, bang") and pressing Enter will execute the last command entered.

If the command is too far back up the list for scrolling to be convenient and you k now the name of the command, you may prefer to jump to it directly, using a s hortcut: ! string (where string represents the command). For example: ! mcopy might find the previously used command: mcopy –tnm something.txt c:/mydir and execute it directly.

If you are not sure of the command name or have used several variations of the command and are not sure which one you want, you can view the entire history file using the command: history (without parameters). The commands will be numbered in the order they were executed. When you find the command you want, you can execute it simply by typing: ! 452 (or whatever the correct number is). If you do not want to view the entire 500 line history, you can elect to see just the last x number of commands, such as: history 50 (to see the last 50).

If you need to search the entire 500 item list but do not want to read every line looking for a command, you can us e the grep command to search the contents. For example, to find that ***mcopy command from earlier, type: history | grep mcopy (where |, the piping symbol, is*** the Shift-Backslash character above the Enter key, on a standard 101-key keyboard). You can even use partial-word searches with grep, such as: history | grep m (to find the ***mformat, mdel and other commands starting with "m" that were executed previously), or even*** wildcard searches (history | grep m*y). If none of these methods finds the command you are looking for, then most likely it was executed more than 500 commands ago and has "fallen out" of the .bash_history file.

Q.38. What can I do if I don't remember the name of a command I need?

Ans. You can page through a manual looking for the command you want, but a quicker way is with the apropos command. It allows you to search a database of available commands, using keywords. For example, if you want to operate a floppy drive but do not remember what command you used the last time (or perhaps this is your first time and you do not k now where to start), type: apropos floppy. You should see a result like:

```
fd (4) - floppy disk device
fdformat (8) - Low-level formats a floppy disk
mbadblocks (1) - tests a floppy disk and marks the bad blocks
in the FAT
mformat (1) - add an MSDOS filesystem to a low-level
formatted floppy disk
mkbootdisk (8) - creates a stand-alone boot floppy for the
running system
setfdprm (8) - sets user-provided floppy disk parameters
```

(The numbers in parentheses indicate the section in the man pages where this information can be found, such as: man mformat to look at section 1 (the default) in the man pages for the mformat command, or: man 8 mkbootdisk to go directly to section 8 to see the rmt command.)

If you were to use the command: apropos diskette instead, Linux might return a "nothing *appropriate" message. This indicates that apropos doesn't recognize the term "diskette." If* this happens on a search, simply try another term to look for, such as "floppy." It is also possible that apropos won't find every possible related command. For instance, if you search for information on tape drives using: apropos tape, it would list a number of commands related to operating the tape drive, including mt and rmt, but it would not include tar, which can be used for tape backup, so you may need to be creative to find what you are looking for.

Note: If the apropos command doesn't work , it may mean that the whatis database (which is accessed by the apropos command) has not yet been built. To create the database, first use the: su – command (and type in your password when prompted) to log in as root, then enter:

/usr/sbin/makewhatis. This process may take several minutes. When it finishes, try the ***apropos command again. For more on whatis, refer to the question, Is there online*** *documentation for most programs?*

Q.39. Just used a command and got completely different results from the last time I used it. Why?

Ans. Remember that Linux is case-sensitive. Verify the case of the parameters you used for the command. For example, if you type: tar –xvf filename.tar one time, and: TAR –XVF FILENAME.TAR on another occasion you will get different res ults, because x and X, v and V, f and F, are all valid, yet different, parameters for the tar command. In fact, because commands are case sensitive as well, you could have two completely different commands, called tar and TAR—with different syntax and parameters—on the same system.

Q.40. Long pathnames are a pain to type in. Is there any sort of shortcut I can use?

Ans. Yes, in some cases. If you are using the Bash shell, or other shell with similar capabilities, you can take advantage of a feature called filename expansion, by using the Tab key. You can type the first part of a directory name, then press Tab and it will complete the name for you (as long as there are no other names starting with the same letters.

For example, if you want to list the /home/johnson/projects/2000 directory, you could type: ls /ho[Tab]/jo[Tab]/pr[Tab]/2 and the shell will expand the pathname for you. But, what happens if you have two directories under /home/johnson, called /projects and **/previews? If you know this ahead of time you could just change /ho/jo/pr/2 to /ho/jo/pro/2,** to eliminate the conflict. Alternatively, if pressing the Tab key after /pr doesn't provide the pathname expansion, pressing Tab a second time will bring up a list of all names that start with "pr". (In other words: ls /ho[Tab]/jo[Tab]/pr[Tab][Tab].) Then simply use the keyboard up-arrow key to repeat the previous command and change "pr" to "pro", or whatever else is needed to create a unique match; then finish the command with the /2 at the end.

Q.41. Is there a way to use one command to start more than one program at a time?

Ans. Yes. The command shell allows you to "stack" commands on one command line, to be executed in order. For example, if you type: ls –l; cd ..; ls –a Linux will list the current directory contents in "long form" (ls –l), then change directory to the parent of the current directory (cd ..), and finally list all files in that directory (ls –a). The "semicolon-space" separating the individual commands is how Linux knows where one command ends and the next begins. If you wanted to copy a file from one directory to another and then to a floppy (while changing its name), and then display the contents of the floppy, you could type:

```
cp /dir1/myfile.dat /dir2; cp /dir1/myfile.dat /dev/fd0/myfile2.dat; ls -
a /dev/fd0
```

Stacking commands can be an efficient way of starting several programs that must run sequentially, without having to stand around waiting for each to finish in turn. Simply type in the stacked commands and walk away until the last one is done. Of course, if you need to perform several tasks end to end on a recurring basis, it makes more sense to create a "script" file (like a batch file, for you DOS/Windows users) containing those same statements. Then you jus t have to type in the name of the script file, instead of a long multi-part command.

> ***Warning: Like many powerful tools, command stacking can be a double-edged sword. It can*** hurt you as easily as help, if you are not careful. As an illustration of what can happen if you are careless, years ago I was using the OS/2 operating system to write a book about OS/2, which also supports command stacking (although without requiring the semi-colon between commands). I decided to delete the existing files from an old disk ette so that I could back up my book chapters (one file per chapter) to it. I intended to type in the following command: del a:*.* which would have deleted all files from the root directory of the a: drive. Unfortunately, what I actually typed was del a: *.* (a space where the backs lash should have been). OS/2 interpreted that as "delete all files from a: (del a:) and then delete all files from the current directory (del *.*)"—all my original files! I realized

what was happening about halfway through the files on the hard drive and was able to cancel the job before the rest were deleted. Fortunately, I had a backup from the night before, so I lost only a few hours work, but it could have been much more serious. The moral of this story is be very careful with command stacking (or script files), because a lot can happen when you run programs unattended. If you were to issue a command to first copy a file to a floppy and then delete the original, and the floppy copy was unsuccessful (the disk was full, or not formatted, or defective), it is possible for the file to be deleted without having been backed up first.

Q.42. Is there a way to stack commands and have them execute concurrently in other command sessions?

Ans. Yes. Running the stacked commands in the examples from the previous question will cause them all to execute one after the other in the same session window, rendering it unavailable until all the commands have finished running. If you would rather the commands launch another command session and run the commands from there, simply append an ampersand (&) symbol to the end of the first command, such as:

```
cp /dir1/myfile.dat /dir2&; cp /dir1/myfile.dat /dev/fd0/myfile2.dat; ls
-a /dev/fd0
```

In this case, Linux would launch a new session for the first command, then continue to execute the remaining commands in that session when the first one is finished. Meanwhile, the original session is available for your use.

If for some reason you want each command to run in its own session, append the ampersand to all of them, as follows:

```
cp /dir1/myfile.dat /dir2&; cp /dir1/myfile.dat /dev/fd0/myfile2.dat&; ls
-a /dev/fd0&
```

Of course, if you are doing this just to free up the original session so you can keep work ing while the other sessions are busy running those commands, a s impler solution might be to just manually launch a second command session to work from while the firs t one is busy.

Launching several programs in different sessions offers possible advantages. If, for example one program is doing heavy database updating while a second is performing spreadsheet recalculations and a third is displaying complex graphic s onscreen, each is primarily stress ing a different part of your computer's hardware. (The database update is using the hard drive heavily, the spreadsheet is grabbing a lot of processor cycles to calculate numbers and the graphics program is mainly using the video controller.) Running them all at once could greatly improve the performance of your computer by using system resources that would otherwise sit idle until the current program finishes.

Bear in mind, however, that how useful this is will depend to a great degree on the kinds of programs you are running and the type of hardware your computer has installed. For example, if you run two programs at once that both make heavy use of your hard drives, people with multiple IDE drives (or one SCSI drive) will likely see little or no performance improvement by doing this. This is because the time spent jumping back and forth between hard drives, or even between areas on the same hard drive, offsets much of the time savings of running the programs together. IDE devices are "dumb," relying on the processor on the IDE controller to handle all the disk drive instructions, but the controller can access only one drive at a time, so it has to stop working on one drive, switch to the other, work for a while, then stop and return to the first. This is very disruptive. On the other hand, for those computers with SCSI drives, the situation is very different. SCSI devices are intelligent (which is why they typically cost so much more than IDE drives) and the SCSI controllers can multiprocess, allowing all SCSI devices to execute concurrently and independently of one another. So if you have two programs accessing data on two different SCSI drives, they can both run full speed all the time. Of course, regardless of disk type, if you do not have enough memory to run both of these programs concurrently, Linux will have to alternate between them in memory, swapping all or part of one program out to the hard disk temporarily. This extra disk activity will reduce the amount of time that can be spent reading/writing the data files, again hurting performance.

Similarly, if one program is making heavy use of the processor and the other is primarily graphics-oriented, there s till may not be as much an advantage as you might think to running them concurrently. If you have a high-end video card with lots of memory, it may be able to process all of the graphics commands us ing the video controller's onboard processor and memory. However, many computers—especially those at the low end of the price range—use inexpensive video controllers with little or no video memory (some share the computer's memory instead, using something called Uniform Memory Architecture, or UMA). This means that the computer's processor may have to process all of the spreadsheet calculations and the graphics calculations at once, reducing the performance of both programs. Moreover, using the main computer memory for both may again result in swapping, unless you have lots of system RAM.

Still, it c an't hurt to try executing your stacked commands concurrently. If they all complete faster than if run sequentially, terrific. If not, you can always go back to running them the "old fashioned" way.

Logging In/Logging Out/Shutting Down : This section contains questions or problems involving shutting down Linux, or logging out of an account.

Q.43. I can't seem to log in, even though I'm using the correct password. What's wrong?

Ans. Linux is case-sensitive, meaning that a password of a1b2c3 is not the same as A1B2C3. Be sure that the Caps Lock key is turned off before typing your password (unless, of course, your password is all caps).

Q.44. Why do I need to use Shutdown? Can't I just turn the power off?

Ans. Just as with Windows, using Shutdown allows the system to save open files, flus h the system cache and do other necessary system maintenance. Turning the power off abruptly circumvents this housekeeping and can result in lost data and other problems. Also, because the file system is left in a "dirty" state, it must be cleaned up the next time the system is booted, slowing the reboot considerably.

Q.45. I can find a Logout option, but where is Shutdown?

Ans. Using the KDE user interface, click on the KDE icon (the large stylized K) on the Panel and select Logout. When the Login panel pops up, click on Shutdown and then one of the three options: Shutdown, Shutdown and restart or Res tart X Server. Shutdown shuts down the operating system and turns the system power off; Shutdown and restart does a restart, and *Restart X Server restarts only the X Server proces s without restarting the system.*

From the Gnome desk top, click on the Gnome ic on (the large stylized G that looks like a funny footprint) on the Panel and select Log out. When the Log Out Confirmation panel pops up, select one of the three choices: Logout, Halt, or Reboot and press the Yes pushbutton. Logout logs you out of your account and brings up the Login panel; Halt logs you out and turns off the system; Reboot logs you out and restarts the system. (Halt and Reboot will require your account password to proceed.) Pressing the No pushbutton will return you to your Gnome session.

From a command line, use one of the following commands: shutdown –r now (to reboot immediately), or shutdown –h now (to halt, or stop, the system). To set an interval before shutting down, replace "now" with "+number", such as: shutdown –r +5 (for a five minute delay). You will be prompted for your account password before the system shuts down.

Q.46. Is there a faster way to "warm-boot" the system than using Shutdown?

Ans. Yes. As with Windows, you can use the Ctrl-Alt-Del key combination to restart the system, but first you must open a full-screen virtual console session. A simple way to do this is by pressing Ctrl-Alt-F1. Once the text mode screen appears you can press Ctrl-Alt-Del. You will *not be asked whether you are sure; reboot is immediate. Be sure that you do not have any* files open before doing this, as you can lose data otherwise. If, after opening a virtual console you change your mind, press Ctrl-Alt-F7 to close the console and return to the graphical desktop.

Q.47. Do I have to use Shutdown if I am just turning over the system to another user?

Ans. No. In this situation, the preference would be to simply us e Logout, allowing the next user to login without the time and effort of rebooting the system.

Q.48. I clicked on Logout and now the session is locked up. How do I log out?

Ans. Just press Ctrl-Alt-Backspace to kill the session and return to the Login screen.

Q.49. How can I enable Linux to automatically restart applications that are running when I use Shutdown or Logout?

Ans. On both the KDE Logout panel and the Gnome Log Out Confirmation panel there is an option to Save current setup. Make sure these options are selected before logging out of these environments.

Files and Directories : The following questions relate to work ing with files and directories.

Q.50. What is the root directory?

Ans. The roots of a tree form the foundation for the entire tree. The trunk of the tree has a few large branches; these branches lead to smaller branches, which in turn have smaller and smaller branches that eventually lead to leaves.

Using this analogy, the root directory is the base (roots and trunk) from which all directories (main branches) and subdirectories (smaller ones) branch-off. Inside these directories are data files and programs (the leaves). This sort of arrangement (used by UNIX/Linux, DOS/ Windows, OS/2 and other operating systems) is often referred to as "tree-structured."

Q.51. Can you explain more how the directory structure of Linux works?

Ans. The main difference between how this works in Linux versus Windows, is that Windows uses backslashes (\) instead of slashes (/) in the directory pathnames. To illustrate, here is the hierarchy of a series of directories branching off the root:

```
/
/home
/home/mark
/home/mark/.netscape
/home/mark/.netscape/cache
/home/mark/StarOffice51
/home/mark/StarOffice51/database
/home/mark/StarOffice51/gallery
/home/mark/StarOffice51/gallery/clipart
/home/susan
/home/susan/.netscape
/home/susan/.netscape/cache
/usr
/ust/etc
```

To show it another way, this is how it looks lining up those same directories by equivalent levels of "depth":

```
/
 /home
 /mark
     /.netscape
          /cache
     /StarOffice51
          /database
```

```
          /gallery
               /clipart
   /susan
       /.netscape
            /cache
/usr
  /etc
```

The / indicates the root, or base, directory. /home is a directory off the root and /home/mark and /home/susan are directories branching off from /home. (These subdirectories hold account-related files for user IDs "mark" and "susan".) Within /home/mark there are directories called .netscape and StarOffice51, which in turn have directories branching off from them. Susan has a similar directory structure for her Netscape data files. Outside of the **/home directory structure, there is also a directory called /usr branching directly off the root.**

To change analogies for a moment, if /mark (and its subdirectories) and /susan (and its subdirectories) are each children (and grandchildren) of the /home directory, then /usr is a sibling (or peer) directory of /home. Both are direct offshoots of the root. (This whole naming convention is referred to as the "file namespace".)

From the root directory, to delve deeper into directories and subdirectories, use the cd (change directory) command. For example, cd /mark/StarOffice51 would immediately move you down the directory path from the root to the contents of the subdirectory **StarOffice51. (Using the command: ls –a at that point would show you the files and** subdirectories stored in StarOffice51.) If you then wanted to change to the cache directory inside of StarOffice 51, you have two ways of doing so. You could type the "absolute" path of: cd /home/mark/.netscape/cache (which says, "starting from the root, travel down these directories") to get there, or you could simply use the shorter "relative" path of: cd cache to get to the same place. A relative path means, "starting from where we are now, go to the next level directory, called cache." (Had there also been a directory below cache, called etc, we could get there from the StarOffice51 directory with the command: cd cache/etc).

Remember, preceding the pathname with a forward slash (/) tells Linux to s tart from the root directory. Therefore a command of: cd /cache/etc would fail in the preceding example (unless you also happen to have a cache directory branching off from the root, with its own **etc subdirectory). Why? Because it is looking for the cache directory starting from the root** rather than from the StarOffice51 directory.

Reversing our direction, if you use the cd .. command from within a directory it will step you back up a directory level (such as from /home/mark/StarOffice51 to /home/mark). You can repeat this command as often as necessary to return to the root directory (/). Alternatively, to jump directly to the root, use the cd / command. (This is equivalent to cd\ or cd \ in DOS/Windows.)

Q.52. There are a number of directories I don't recognize in the root directory. What are they?

Ans. These directories store essential system files and programs. For example, the /etc (et cetera) directory contains configuration files, where system and application settings are stored. Beneath that, the /etc/skel directory contains shell, or "skeleton" files used to create user accounts. The /usr (user) directory stores other important files, including program libraries (shareable system program modules) in /usr/lib, manual pages for the man utility in /usr/man and other documentation in the /usr/doc and /usr/info directories. Other such system directories include /lib (libraries: security files, system modules and libraries containing software libraries and informational databases), /opt (optional: pre-loaded software packages are often stored here) and /var (variables: s pooling, logs and all messaging files). Unless you have a specific reason, and k now what you are doing, it is a good idea to leave those directories alone.

Q.53. Is there a command line shortcut to the login directory?

Ans. Yes. From any shell prompt simply type: cd and press the Enter key.

Q.54. Some filenames are preceded by a dot. What does this mean?

Ans. A filename that starts with a dot—such as .bashrc or .netscape—is a hidden file. They are often used to hold user preference settings, and other configuration information. Thes e files are hidden to make it less likely that they will be accidentally deleted. Hiding files also makes a file listing s horter and less cluttered with files the user will rarely be interested in. There are different ways to see these files. For example, the ls (list) command with no parameters won't display these files, but ls –a (list all) will.

On the other hand, if you are using the K File Manager (KFM) under KDE, simply select View, then Show Hidden Files from the pull-down menu to see these files. (To start KFM, just click on the Panel icon that looks like a house in front of a manila folder.)

Gnome users can click on the Gnome icon (the funny G-shaped footprint) on the Panel, select *Programs, then File Manager. From the Gnome File Manager, choose Settings, then Preferences; finally, clic k on Show hidden files.*

Q.55. When I look at a list of files and directories, the names are followed by something like -rw-rw-r— or lrwxrwxrwx. What does this mean?

Ans. Thes e cryptic notations actually describe the "permissions," or access levels, assigned to three groups: the file owner, groups and other users. The first position identifies what type of object the item is . Possible types include a physical (actual) Directory/folder (d), a physical file (-), a mere Link /alias (pointer) to a file or folder (l)—similar to a Windows shortcut, a Sock et (s), a pipe (=), or a device (b is for "Block" devices, such as disk drives, that handle data a block at a time; and c is for "Character" devices, like serial and parallel ports, that deal with data one byte, or character, at a time). The remaining nine positions are composed of three groups of three characters each.

The first three apply to the file owner (usually its creator). The next trio describes the permissions granted to any authorized groups of users and the final three characters denote any access available to unspecified other users. Each triad of positions reveals whether the user has Read (r), Write (w)—i.e., create/modify/delete, or Execute/run (x) permission, or no permission to perform one of those actions (-).

Using the first example from the question, -rw-rw-r— tells us that the object is a file (because the first character is -). The first rw- indicates that the owner has read/write access to the file, but not execute (so it can't be a program, otherwise the owner would be able to run it). Likewise, any groups that have been granted access to the file can Read and Write to it (the second rw-). However, any "other" users (those not expressly added to the authorized group) have only Read access to the file, as evidenced by the r— at the end. If the permissions were **drw-rw-r— instead, the only differenc e would be that the object is a folder/directory rather than** a file.

In the other example from the question, lrwxrwxrwx means that the object is a link to a file or folder and all three groups have read, write and ex ecute permission to access that object. Execute permission means that if it is a program, script or other executable file, an authorized user (in this case, anyone) c an run it, and if it is a folder then the user can read the contents.

Q.56. The ls command doesn't show everything in the directory. What's wrong?

Ans. The ls (lowercase L, not capital i) utility has many parameters to let you view the directory contents in different ways. For example, ls –a shows all files and ls –al shows all files in "long" form (with additional details), while ls –a —color adds a splash of color to differentiate the various types of files. (For the list of parameters, use: ls —help or type: man ls for a full description.)

Q.57. Can I use wildcard searches with the ls command, as I can with Dir in DOS/Windows?

Ans. Sure. You can search for all files ending with the .txt extension using: ls *.txt, or all files starting with "mc" and with any extension, using: ls mc*.*. If you are looking for a file among a long list of files with similar names, you might limit the wildcard to just one character, using the ? symbol. For example: ls hello?5.c would find files called hello05.c, hello95.c, **helloj5.c and so on, but not hello105.c or hellor7q5.c. You can even combine wildcard** characters for very specific searches, such as: ls hello?5.*. (Wildcards can also be used with other commands, including cp (copy), mv (move), rm (remove), etc.)

Q.58. I want to do a wildcard search on files with ? or * in the names. How do I do that?

Ans. Unlike DOS/Windows, UNIX/Linux does allow those characters, as well as) and (and | and other special characters in file and direc tory names. To distinguish between a wildcard search using ? or * and a search for those particular characters, you merely need to insert a backslash (\) before the wildcard character. For example: ls hello\?.txt or ls myfile*.c or ls mc\)127.h would find the files hello?.txt, myfile*.c and mc)127.h, respectively. You can also combine thes e "regular expressions" with wildcards in searches, such as ls hello\?.t?t (which would list hello?.txt and hello?.tst), or ls mc\)127.* (which would list mc)127.h and mc)127.c, for example).

Q.59. How do I create or rename a file with special characters in the name?

Ans. As with the previous question, you merely need to insert a backslash before the special character. For example: touch hello\?.txt would produce an empty file called hello?.txt; touch myfile*.c would result in myfile*.c; and touch mc\)127.h would create **mc)127.h. Similarly, mv hello.txt hello\?.txt would add a ? to the filename,** immediately before the dot. To use multiple special characters, simply insert a \ before each one; for instance: touch myfile\(**\).txt would produce myfile(**).txt. The procedure is the same for any other special character. If you are not sure which characters require the \ treatment, simply try the command without it first. If you receive a "sh: syntax *error near unexpected token 'yourfilename.'" message (where 'yourfilename' is the name you* tried to create), then you need the backslash.

> ***Important Note: Although you can include many special characters in this manner, using*** some special characters , including the asterisk (*), question mark (?), hyphen (-), vertical *bar/piping symbol (|), circumflex or caret (^), apostrophe (') and ampersand (&, in a file name is generally not a good idea, because they may be confused with wildcard, parameter*("argument") and other symbols by various commands. In addition, using a space in a filename (such as my file.txt) may cause some commands to treat the name as two filenames, or a parameter and a filename. (To avoid this problem, enclose the name in quotes, as in "my file.txt".) Any of these conditions will cause unintended results. At the very least, these special symbols can make filenames more difficult to read. Unless you have a good reason for doing so, it is probably best not to use them.

Q.60. How long can a filename be?

Ans. A filename can be up to 255 characters long, including numbers, letters and special symbols, as well as a combination of upper and lower case (which are treated as different characters by Linux/Unix). However, for the sake of readability and to make it easier to type from the command line, much shorter names are preferable. This 255-character limit does not include the full path. In other words, if a file called ThisIsMyVeryLongFilename.txt is in the **/user/mark/garbage/files directory, the filename alone can be up to 255 characters,** excluding the charac ters in the pathname. To illustrate, out of the following path:

```
/user/mark/garbage/files/ThisIsMyVeryLongFilename.txt
```

only the 28 characters (bolded) following the final / comprise the actual filename (including the dot before "txt").

Q.61. The directory listing is too long and scrolls off the page. Is there a way to view a page at a time?

Ans. Yes. There are programs, called "pagers" to do this. Just as in DOS/Windows, where you could use: dir c:\ | more to lis t the contents of the root directory, in Linux you can type:

```
ls -a / | more
```

Or, using an alternative utility:

```
ls -a / | less
```

Using less, press the Spacebar to move ahead a page at a time; to back up, use the B key. When you are done, press Q to quit.

Q.62. Is there a way to find files from a command shell?

Ans. Just as Windows has a Find feature, Linux has the locate command. It will find data files, programs, directories and other objects whose names match your search argument. For example, locate myfile will find myfile.txt, /myfile and other matches. (For more on ***locate, including search parameters , use: man locate.) Locate searches a database (called slocate.db) on your hard drive used for this purpose, which is updated overnight. If you are*** searching for a new file that locate is unable to find (or you regularly s hut the system down overnight, so the database doesn't get updated often), you can force a manual database update. To do this, use the: su command to temporarily login as root, then from the shell prompt type: updatedb. Wait a few minutes for the database update to complete and try the ***locate command again.***

Note: Linux does also have a find command, but for performance reasons locate is better for searching on filenames. If you have a need to search by file date or other attributes or would like to automatically execute a program us ing the found file as a parameter, or any of a large number of other interesting uses, find is the command to us e. For more on find, use: info find.

As an alternative, if you are merely looking for the binary, source and manual pages for a command, you might try the whereis command. To illustrate the difference between whereis and locate try the following two commands: whereis passwd and locate passwd. As you can see, the whereis command found only a few files, while the locate command listed dozens. That is because whereis was look ing for a few specific files (all called passwd.*), while locate was busy seeking any file or directory name containing "passwd" in the name (including gpasswd.1.gz, autopasswd, passwd-1.html, etc.) For another example, try: whereis *.doc and locate *.doc. As before the locate command came up with many files sporting a .doc extension. The whereis command, however, found nothing. Why? Because there is no Linux command called "*.doc" or even "doc". Remember, whereis works *only on files associated with commands, not just any old data file or program.*

Q.63. Is there a way to browse through a text file from the command line?

Ans. Yes. Perhaps the best way is to open the file with a text editor, such as Pico, Emacs, Vi or others, and scroll or page through the file. However, if you are trying to do the browsing from the command line without opening the file first, you can use the head and tail utilities, or the ***more or less commands. For example, head myfile.txt would display the first 10 lines of*** the file onscreen, by default. The command: head 40 myfile.txt would show the first 40 lines. (You can substitute larger or smaller numbers for "40".) Similarly, tail displays the last 10 lines of a file, as in: tail myfile.txt. (If the file is short, but you are not sure how many lines long, you could use the cat command (short for concatenate) to display the entire file on the screen, such as: cat myfile.txt.)

Obviously, this is fine if you k now approximately how close to the beginning or end of a file the text string is, but not if it is somewhere in the middle of a large file. In that situation more or ***less would be more productive. (See the question, The directory listing is too long and scrolls*** *off the page. Is there a way to view a page at a time? in this section, for information on those* commands.)

Q.64. Is there a command to search text files for a specific character string?

Ans. Yes. Naturally, you can open a file with a text editor and use its search function. However, if you prefer to search from the command line, you can use the grep utility. It is a very powerful text file search tool. You can perform wildcard searches on a specific file, but more importantly, you can search many files at once, also with wildcards. For example, say you have a directory full of text files and you know that one of them contains the phrase, "Mary had a little lamb", but you do not know which file it is. Using the command: grep "Mary had a little lamb" *.txt might result in the following output:

```
myfile.txt: Mary had a little lamb. Its fleece was white as
snow. And everywhere that Mary
```

(Note the use of quotation marks around the phrase to identify the search string when multiple words are involved.) The grep command shows the entire line of text containing the search string to make it easier for you to determine whether it is the correct tex t (because there may be multiple matches).

Note: The grep utility is by default case-sensitive, so searching for "Mary" would not find mary or MARY. If you wish to perform a case-ins ensitive search, add the –i parameter in the preceding example. Then it will find Mary, mary, MARY, mARy, or any other mixture of case.

For example, a case-insensitive wildcard search such as: grep –i mar* *.txt might yield this result instead:

```
myfile.txt: Mary had a little lamb. Its fleece was white as
snow. And everywhere that Mary
personnel.txt: Candidates for a pay increase this year include
Johnson, Martin and Littlefield.
personnel.txt: Due to Smith's impending marriage, she will be
taking a two-week honeymoon in
```

Not only were there "hits" in two different files, but there were two separate occurrences of "mar" in the personnel.txt file as well.

The results of a search can even be copied to another text file or printed, using the piping symbol (|). (See the question, The directory listing is too long and scrolls off the page. Is there *a way to view a page at a time? in this section, for information on piping.) Search results can* also be redirected to a file using the redirection and append symbols (> and >>). (See the question, Is there online documentation for most programs? in the General Usage section, for information on redirection.)

There is much more that can be done with the grep command. For example, the –c parameter tells grep to display only the count of the number of times a string appears in a file, rather than the text of the matches (such as personnel.txt:2). For more information, read the manual page (command: man grep).

Q.65. What are "symbolic" and hard links, and how can I identify them?

Ans. Symbolic (soft) links work much like "shortcuts" in Windows, as pointers to programs, data files or directories located elsewhere. Links allow you to start a program from more than one folder (or the Desktop) without having to duplicate the files in multiple places. Links are independent files—only a few bytes in size—that can be renamed or moved without affecting the original programs or files they point to. Conversely, if the original file is moved, deleted or renamed it will break the link , because it no longer matches the filename and location pointer stored in the link. (If a link is broken this way, it s hould be deleted—as it is no longer useful—and a new link created, if needed, for the new filename and/or location. On the other hand, if you rename the original file back to what the link is expecting, or return it to its original location, the link will "magically" work again.)

Symbolic links can be identified, using the K File Manager (k fm) or other graphical file manager, by a small arrow in the lower right corner of an icon, pointing at the icon. From the command line, one command you can use is: ls –F (capital F, not lowercase). When the filenames are listed, the symbolic links will be followed by an @ symbol. For example, if you were to list the files in the /bin directory, you might see:

```
arch*
ash*
```

```
ash.static*
aumix-minimal*
awk@
basename*
bash*
bash2*
bsh@
```

and so on, where awk@ and bsh@ are the symbolic links to files in other directories. (The * following the other names indicates that they are "original" files.) Alternatively, you could use the: ls -il command (see below).

Link files specific to the KDE user interface will have .kdelnk filename extensions, such as **Autorun.kdelnk.**

Another type of link is the "hard" link . This is similar in function to a symbolic, or soft, link except that renaming, deleting or moving the original file won't break the link (because it actually links to the physical file on disk, not just the file pathname in the dis k directory structure). Hard links apply only to data files or programs, though, not to directories.

All files have at least one hard link (the actual filename in the directory listing). Adding another hard link increments the hard link count, which you can display using the: ls –il command.

For example:

```
619880 -rw-rw-rw- 2 ownerid groupid 428 Apr 14 19:42 myfile.txt
145389 -rw-rw-r– 1 ownerid groupid 7285 Sep 30 10:17 coffee.txt
268394 -rw-rw-rw- 1 ownerid groupid 8912 Nov 4 19:42 tea.txt
619880 -rw-rw-rw- 2 ownerid groupid 428 Apr 14 19:42 person.txt
```

(The –i parameter says to display the "inode" number—explained below, and the l argument tells Linux to show the directory list in long form, including the permissions, owner and group Ids, file sizes, and creation date and times.) There are four different filenames listed, but notice some key similarities between two of them: the files myfile.txt and person.txt have exactly the same date, time, file size and properties. In addition, they both show "2" for the number of hard links, but all of these similarities could just be a coincidence (after all, two unrelated files of the same size and type could have been created within a minute of one another). The real clincher, however is in the set of numbers preceding the file permissions listings. (For an explanation of file/directory permissions, see the question, How do I display *the permissions for a file or directory?) The "619880" number is the inode, or physical disk* address of the file. Only one file c an occupy a given block of disk storage (just as only one car can occupy a particular parking space at any one time). Therefore, thes e "two" files must in fact be only one physical file with two hard links (filenames) to the file.

Again, unlike soft (symbolic) links, deleting either of these two files doesn't affect the other. Another difference between the two types of links is that hard links can't span disk drives. In other words, a hard link can't be created on /dev/hdb to link to a file on /dev/hda.

Q.66. How do I create links (symbolic or hard)?

Ans. The ln (link) command can be used to create both types of link s. For example, to create a hard link called myfile2.txt to a file called myfile.txt, type: ln myfile.txt myfile2.txt.

This will create the link in the same directory as the original file. To create a symbolic link called myfile3.txt to the same original file, use: ln –s myfile.txt myfile3.txt (the –s parameter is what tells Linux to make the link symbolic).

If you use the: ls -il command to list the directory contents, you should see something like the following:

```
619880 -rw-rw-rw- 2 ownerid groupid 428 Apr 14 19:42 myfile.txt
619880 -rw-rw-rw- 2 ownerid groupid 428 Oct 2 10:18 myfile2.txt
224516 lrwxrwxrwx 1 ownerid groupid 6 Oct 2 10:18 myfile3.txt -> myfile
```

Note that all details of the hard link (myfile2.txt), including the inode address (619880), are identical to that of the original file (myfile.txt) except for the creation date and time. It is treated as a physical file (as verified by the file size and the permission type of -), but at the same location as the original file (same inode address) with the same read/write permissions. The 2 following the permissions reflects the number of hard links to the physical file (the directory listing for the original file being the first hard link). Conversely, virtually every aspect of the symbolic link (myfile2.txt) is different. The inode address differs because there is a physical file created elsewhere on the disk that contains a pointer to the original file. Likewise the permissions differ, with the symbolic link showing a file type of link (the 1 leading off the permissions) and including execute permissions for all user types. The 1 following the permissions demonstrates that this is not a hard link. Finally, the filename is followed by "-> **myfile", which reveals the name of the original file to which it points.**

Note that creating a hard link for this file in the same directory probably would not be the normal usage of a hard link. More likely, you would create it in another directory so that you could access the file from either place (or several places, if you make multiple links). For example: ln myfile.txt /user/mark/myfile2.txt would put the new link in the **/user/mark directory.**

Linux itself makes use of hard links. In many cases Linux continues to use programs and files that were first used in earlier versions of Linux, which can sometimes cause problems. For example, if the X Window System was first provided in a directory called /usr/X11R4, Linux software which used X Windows would look for those files in that directory. But when a newer version of Linux comes out that uses a later version of X Windows, the files might be stored in **/usr/X11R5 or /usr/X11R7 or /usr/X11R10. So how to keep from breaking all those programs** that are expecting the files to be stored in /usr/X11R4? Simple. Put hard links from one directory into the other. This way, whic hever directory the application software looks in for those files, it will find what it needs.

As an alternative to the command line interface, some graphical file managers (such as GNU Midnight Commander), offer options to create either or both types of links by right-clicking on a file icon.

Q.67. Can you summarize the differences between symbolic and hard links and differentiate between using links and simply duplicating a file?

Ans. Sure. If you merely copy a file from one directory to another, you have doubled the disk space used by the file, and editing one file doesn't affect the other (because they are in no way connected to one another). This is fine if it is what you intended (so that you can keep a pristine original while modifying another copy, or if you made a back up copy for safekeeping.) If you copy a directory instead of a file, you also copy the entire directory's contents, which can result in a lot of disk space used. A file or directory copy can be put anywhere on any disk, either directly attached to the system or on a network or Internet/Intranet drive somewhere else entirely. Because the original and the copy are unconnected, deleting one has no effect on the other.

On the other hand, if you create a symbolic link it takes only a few bytes of space and can be located on any mounted device that you can type in a path for. If you delete, rename or move the original, you break the link (although recreating the original in the same place will restore the link). Using the link to access the file or directory, any changes you make to the file or to the directory contents will affect the original files/directories, because those are what you are actually touching.

Alternatively, if you create a hard link instead, no additional disk space is used, however the link can only be created on the same physical hard disk (but multiple partitions on that disk), not on other local or remote devices). Moving or renaming the original file won't hurt the link; it will reflect the change. When you "delete" the original file, you are actually deleting only the original hard link (the directory name). So, as long as at least one other hard link exists, the file is still accessible. Hard links can be created for files but not for directories. Because the permissions indicate that a hard link is a physical file, any software that tries to access these files will think it is accessing the original files (which, in fact, it is—through the link).

Q.68. I recognize a number of file types, such as .JPG, .GIF, .WAV, .TXT, .HTM and .ZIP from Windows, but I see many other file types I don't know. What are they for?

Ans. There are a number of file extensions that are common in Linux/UNIX but not typically seen in the Windows world. Here are some of them:

.a – Archive file	.png – Image file (like .jpg/.gif)
.au – Audio file (lik e .wav)	.ps – PostScript printer-formatted file
.bz ip2, .bz2 – Archive file compressed with *bzip2*	.rpm – RPM Pac kage Manager file
.c – C programming language source file	.so – Program library file
.conf – Configuration file	.tar – Archive file
.cpp – C++ language source file .tar.bzip2, .tar.bz2 – Archive file compressed with bzip2	
.gz , .gzip – File compressed with gzip .tar.gz, .tar.	Z – Archive file compressed with *gzip*
.h – C or C++ language header file .tga – Image file	
.html – HTML file (same as .htm) .tgz – Archive file compressed with gzip	
.o – Program object file	.xpm – Image file
.pl – Perl script file	.z, .Z – Older gzip file

If you encounter a file type you do not recognize (such as one with no file extension), you can use the file command to find out more about it; for example: file smith.xyz or file smith (if there is no file extension).

Q.69. Should I back up my entire system?

Ans. That is not as easy a question to answer as it may sound. A fully configured system with operating system, many large applications, data files and perhaps a library of multimedia files (music, video, photographs, etc.) can amount to many gigabytes of information. Unless you have a high-capacity, high-speed backup option (such as a 4mm, 8mm or DLT tape drive, or an Iomega Jaz removable disk cartridge, for example), it just is not practical to back up everything. On the other hand, because you probably installed all your software from a few CDs, it is not absolutely necessary to back up the operating system and applications. If you back up only your data files, setup files, account files and others that have been created or modified since installation, you c an always reinstall the software from CD, then restore the backed up user files.

Of course, the disadvantage to this, in the event of catastrophic disk failure, is the time and effort needed to install all that software individually, rather than just starting a restore program, inserting a tape or Jaz cartridge and walking away. But if you do not have a high-capacity backup solution at your fingertips, a partial backup is the most effective method for all but a complete system rebuild. Each application will have its own set of configuration files that should be backed up, as well as data files, but you will have to find out what they are from the software documentation or the vendors.

From a Linux standpoint, the directories you s hould be sure to back up include:

- /etc
- /home
- /opt
- /usr/lib
- /usr/local

- /usr/src/linux (if you have modified any of the kernel programming)
- /usr/X11R6/lib/X11 (if you are using X Windows)—Case is important in this folder name!
- /var/spool/mail

Of course, you should also back up any directories containing important user data files. If you lack large-capacity storage, you can always back up each application's files to its own set of floppies using an archiving/compression utility, such as tar, gzip, bzip2 or zip. These and other tools can span more than one disk or tape to store all the data in one related archive. This method may not be as quick or efficient as a tape drive, but at least you will have a copy of your essential files. It is not necessary to back up the program files themselves, as you can always just reinstall the applications from scratch, but you might want to back up the files that contain the user setup information (preferences), such as default margins and tabs for a word processor. (One clue as to whic h files these are is to look at the file dates. Generally, the user setup files will have much newer dates than the files created when the program was first installed. If in doubt about what to back up, contact the software vendor for the filenames.)

Q.70. How often should I back up my files?

Ans. There is no one right answer to this question either. Every user is different. One way to answer the question for yourself is to use the "ouch" test. Consider how much work you could afford to lose before it hurts. If you are a home user, perhaps a once-a-week backup would be adequate (right after you pay bills, for example). In a high-performance office environment, once or twice a day—even hourly—may be required.

It is also a good idea to use a "grandparent-parent-child" strategy of rotating backup media. (In other words, use three or more sets of tapes or cartridges and alternate among them.) This way, if for some reason your most recent backup is corrupted, at least you have the next older one to fall back on. Plus, if you discover that you accidentally deleted a file a few days ago—or it was damaged by a virus—you can restore the file from a backup made before the deletion.

Another option is to back up only what has changed since the last back up (some archival programs track this information from backup to backup). The advantage is that often very little has changed since the last time, making for very quick back ups. The disadvantage is that you must keep copies of all the intervening backups since the last complete archive. Losing even one partial backup along the way can result in the loss of important information. Thus, even with a policy of incremental backups, it is a good idea to do full backups periodically.

Q.71. What backup software should I use?

Ans. There are a large number of commercial and free programs available for use, some primarily for tape drives and others for more general usage (magnetic and optical disks, network drives, etc). Many of these have graphical interfaces to make them easier to use. It is a good idea to try out a few different programs to see what you like best.

There are too many available tools to try to describe them all here. But there are a few common commands and utilities—likely to ship with your Linux distribution—that you can start with:

tar — tar was the earliest and most common "packaging" tool for Unix/Linux (it does the *unpacking of files as well—there is no "untar" command). The tar command creates a file that* contains all the other files and directories, optionally in a compressed state (us ing gzip or ***bzip2), to save space. If you have used zip/unzip or PKzip with DOS/Windows, then you are*** already familiar with the concept. Files pack aged with tar typically end with a .tar extension. If also compressed with gzip, the extension may be .tar.gz, .tar.Z or .tgz. Tar files compressed with bzip2 generally have a .tar.bz2, .tar.bz, .tbz2 or .tbz extension. The gzip and bzip2 compression utilities can also be used independently of the tar command to compress and decompress files (see below). There are many different parameters available for the tar command (whic h does the unpacking of files as well), so it is best to read the manual page for ***tar using the command: info tar. Note: It is not advisable to c ompress tar files on a tape*** backup, because tar would not be able to recover from tape errors.

gzip/gunzip — gzip (GNU zip) is the tool traditionally used for compressing tarred files and its companion gunzip is used to decompress those files . More recent tools, such as PKunzip, can generally uncompress files created by gzip as well as those compressed by other tools, such as PKzip and zip. Due to this flexibility and because PKzip and zip are also available on other operating system platforms, they have become more popular than gzip. Files compressed with gzip usually end with .gz, or in the case of s ome older files, .z or .Z. You can read more about the gzip/gunzip commands using: info gzip.

bzip2/bunzip2 — Think of bzip2 as a newer, fas ter version of gzip. It has a very similar command structure (including many of the same parameters), but it tends to be faster at compressing and decompressing than gzip/gunzip. Bzipped files end with a .bz2 or .bz extension. Learn more about bzip2/bunzip2 from: info bzip2.

PKzip/PKunzip — PKzip is a popular shareware/commercial packaging and compression tool, primarily in the DOS/Windows world. PKzip is very fast and offers varying degrees of compression vs. speed. Newer versions include a graphical front user interface. PKunzip is compatible with files created by PKzip, zip, gzip and tar, as well as the UUencode, XXencode, MIME and BinHex file formats. PKzip files end with a .zip extension. For more information, read the documentation that comes with PKzip.

zip/unzip — zip and unzip are a pair of command-line tools that will package and compress (and decompress) files and directories . In addition to being very fast, they are free and available for other platforms, including DOS/Windows, OS/2 and other UNIX OSes, including SCO. The unzip command can also unpack files created by PKzip. Lik e PKzip, zip-compressed files end with a .zip extension. The zip and unzip utilities may be included with your Linux distribution; if so, read their manuals using the following commands: info zip and info unzip.

Other utilities you can obtain separately include Arkeia, Backup Edge, BRU2000, Catchup, CTAR, **Drive Image, Kbackup, KDat, PerfectBACKUP+, Sitback, Taper and many more.** For a lot more on this topic, see the related white paper entitled, Installing Linux Applications *for the First Time, available from the same sources as this paper.*

Root/User Accounts, Groups and Permissions : Thes e are questions revolving around administering accounts, passwords and permissions.

Q.72. Sometimes instructions say that I must "be root." What does this mean?

Ans. It means that you must be logged in as the root user, or superuser. (Sorry, but you do not get a cape with a big S on it!) There are two ways to do this. If you will be working as root extensively, you should logout as yourself and login as root. When you are done, reverse the process and log back in as yourself.

On the other hand, if you need to be root only long enough to execute a command, you can open a command shell and issue the command: su (substitute user) to temporarily login as another user, while still inside "your" login shell. You will be prompted for the root password. Once that is provided you will "be root" and can issue commands as needed until you end the root session by typing: exit and pressing Enter or by closing the shell.

Some commands won't work unless you have logged in as root (i.e., su won't work, because Linux "knows" it is still you in charge). An alternative is to use the su command (su followed by a hyphen, with a space separating them). This makes you root with the root's login shell, as if you had logged off and logged back in as root.

Q.73. What is the root account and how does it differ from a user account?

Ans. When you installed Linux, you were asked to provide a root password. Later you were prompted for a user account name and password. The root account is equivalent to the system-level access you are given under Windows, with the ability to delete, create and format partitions , create/delete directories and files, or do anything else your heart desires . Linux operates like a network client/server system. The root account is like a system administrator, able to control what access authority the various clients (users) have. There can be many different user accounts set up on the same Linux system, each with its own set of permissions, configuration options, choice of user interface and so on.

This is very useful in a business environment where multiple people share the same system, as each user account can be tailored for each individual user. On the other hand, it can be a nuisance when there is only one user, because there are some system functions that can be performed only by the root account, necessitating that the user log out of the user account and log in to the root account long enough to perform the needed operation, then log out of the root account and back into the user account. (Fortunately there is a shortcut you can use in some situations as an alternative. See the description of the su - command in the earlier question Sometimes instructions say that I must 'be root'. What does this mean?.)

As compensation, however, doing things this way makes it much less likely that the user will accidentally delete or corrupt important system files that they have no business playing with in the first place.

Important Note: Most, if not all, Linux distributions require that you create a user account in addition to the root account when you install Linux . While it is tempting to log on as root and use the root account for everything, this is a Very Bad Idea. There are too many ways in which you can accidentally (or someone else can maliciously) corrupt up a system from the root account. By running mostly from the user account, your system is protected to a great degree from most disasters. (It won't prevent a disk crash, of course, but at least you can't accidentally delete your /usr/lib directory, for example.) If the Linux distribution you installed required you to create a user ac count, use it. If you did not create one during installation, create one now—and use it!

Q.74. Why would I want to add other user accounts to my system?

Ans. If you are the only one who uses your system, there may be no need to set up other user accounts (besides your own). On the other hand, if you share your computer with other people (second and third shift workers , perhaps; or you share a common work area; or other members of your family use your computer), or even if you wish to make your printer or other devices available to others on the same network, you will need to set up individual user accounts and possibly group accounts.

Q.75. How do I add users?

Ans. There are a number of tools available for doing this. For instance, Red Hat Linux 6.2 includes the graphical Red Hat Disk Management tool (linuxconf). For occasions where you are not running a graphical interface, or prefer a command line tool, a couple of text-mode programs that are often included with Linux are adduser and useradd. Despite the similarity of their names and purpose, adduser is the more functional of the two.

To use linuxconf:

1. Run: linuxconf from a command line under either KDE or Gnome. (Linuxconf is not located on a menu anywhere.)

   In order to use Linuxconf and similar system-level tools, you will need to first be logged in as root. To do this, use the su – command and enter the root password when prompted.
2. Once the linuxconf panel opens, scroll down to the heading called Users accounts. Immediately below, there is a subheading called Normal, and beneath it User accounts.
3. Double-click on User accounts to open a panel that lists all existing accounts.
4. To create a new account, click on the Add button at the bottom of the panel. You will see a s ubpanel with four tabs: Base info, Params, Mail settings and Privileges.
5. When you are done filling in the necessary information, press the Accept button to save your changes (or Cancel to reject the changes and not create a new account).S

Users of other distributions besides Red Hat have other graphical user administration tools at their disposal. For instance, Caldera OpenLinux 2.3 users have Webmin and COAS.

Although neither individually can do everything linuxconf c an do, together they offer similar functionality.

To add users with the COAS tool:

1. From the KDE menu, select Settings, then COAS, then System and finally Accounts. A panel will appear.
2. On the tool bar, s elect Us er and then Create user.
3. When the menu appears, begin customizing as many users as you need.

If your distribution does n't offer a graphical tool, or if you prefer to use the command line interface, you can use adduser. As with linuxconf, before using adduser you must login as root. Then type: adduser newusername (where newusername is the user name you wish to create).

The program will assign the next available ID for both the user and default group, then remind you to set a password for this new account. (To add the password, see below.) This will supply the minimum information necessary to create a user. To add personal information to the account record, such as the user's name, address and phone number, use the chfn (Change Finger Name) command (see below).

Q.76. How do I delete users?

Ans. This is a multistep process:

1. Log in as root.
2. Use the userdel command with the name of the user you wish to delete, as in: userdel mary. Note, however, that while this deletes the user account name from the /etc/passwd file, it doesn't delete all associated files.
3. Follow up with the groupdel command to remove the user's UPG entry (the default group), such as: groupdel groupname.
4. Remove any other group affiliations for the user (besides the UPG), using: usermod –G username username. (Note that the –G parameter is capitalized.)
5. Use a text editor to remove the login ID from the /etc/group file.
6. Delete the user's home directory and mailbox using: rm –R /home/username /usr/spool/mail/username. (Note that –R is capitalized.)

Q.77. Is there any way to change a user ID?

Ans. Yes. To change a user ID (login name), use the usermod command. For example, to change an account name from "mary" to "Mary_Jones ", use: usermod –l Mary_Jones mary. If you change a login name, you should also change the user home directory, where all the user account information is stored. To do this, use the command: usermod –d /home/Mary_Jones –m mary to move the user account information to the directory**/home/Mary_Jones from /home/mary. (The –l parameter tells usermod that you are** changing the login name; –d identifies the directory you are changing; and –m means to move the contents to the specified direc tory.)

Q.78. Is there a way to disable a user account?

Ans. Yes. If you are using "shadow" passwords (see the question What is a shadow password?, in this section, for an explanation), the simplest way to disable an account is via the usermod command. Merely assign an expiration date that has already passed, such as: usermod –e 01/01/92 mary. (See below for an explanation of shadow passwords.)

As an alternative, experienced users can edit the /etc/passwd file. To disable the account, find the line for the appropriate user and insert an as terisk (*) in front of the encrypted password. The password is located between the first and second colon separators, following the user name. For example:

```
mary:*Ms.2jbeTWmUYi:500:Mary Jones:/home/mary:/bin/bash
```

Warning: This alternative approach should be used only as a last resort, by a proficient Linux user, because manually editing this file can cause major problems if not done correctly!

Q.79. How do I add or change personal information in an account?

Ans. If you wish to add name, address and telephone number information to your account or change information already there, there is the chfn (Change Finger Name) command. You will be prompted for your password, then shown the data fields for the information above.

To change other user accounts, login as root. Then specify which user you wish to change the information for, such as: chfn fred, and make the additions or changes.

If you prefer a graphical tool, Red Hat Linux 6.2 users running KDE can click on the KDE menu (the big K on the Panel) and select the Red Hat option, then System and finally About *Myself. This will bring up a panel with entry fields for all the information accepted by the chfn* command. I am not aware of any such graphical tools for desktop environments other than KDE, so if you are not using KDE you will have to use the chfn command line option.

Q.80. How do I add groups?

Ans. Default groups are automatically created when you add users. Should you need to add groups manually you can do so with the groupadd command. Normally, group IDs (GIDs) are automatically supplied (by adding 1 to the last number used) when a group is created. However, because you are creating a group manually you will need to specify a group ID yourself. The maximum GID number is 65536, so you might want to specify something very large (say, 50000 or higher) to keep from possibly running into one already assigned by the system. To create a group called "mygroup" with a GID of 50000, type: groupadd –g 50000 mygroup.

Q.81. How do I add/delete users in a group?

Ans. To add users, use the passwd program with the –a (add) parameter. For example, to add user "mary" to group "mygroup" type: gpasswd –a mary mygroup.

To delete users, simply replace the –a parameter with the –d (delete) parameter, such as: gpasswd –d mary mygroup.

Q.82. How do I change a password?

Ans. To change your own account password, Red Hat 6.2 users running KDE can click on the KDE menu and select Red Hat, then System and finally Change Password. This will bring up a panel prompting you for the current password (for the logged in user). If the correct password is supplied, you will be prompted for a new password to use. (If it is too short or invalid for some other reason, the prompt panel will reappear.) I am not aware of any graphical password management tools for desktop environments other than KDE, s o if you are not using KDE you will have to do it from the command line.

From the command line, simply use the passwd command. It will prompt you for your current password and then the new password twice, for verification.

To change another password, you must be logged in as root. Then tell the program which user you wish to change the password for, such as: passwd fred. You won't be required to know the current password; simply enter the new pas sword twice.

Q.83. What is a shadow password?

Ans. Shadow passwords are used to protect system passwords (for user accounts) by making the file containing those passwords (/etc/shadow) readable by the root operator only. When shadow passwords are used they replace the encrypted password in the /etc/passwd file with asterisks. (Moving the passwords to /etc/shadow makes it less likely that the encrypted password can be decrypted, because only the root operator has access to the file.) To see if you have shadow passwords enabled on your system, use the command: ls /etc/shadow. If you receive a message like: ls /etc/shadow: No such file or directory, the files have not yet been installed. For more information on shadow passwords, read the manual page: man 5 shadow.

Q.84. How do I display the permissions for a file or directory?

Ans. From a command line, use the ls command with the –l parameter. For ex ample, ls –a –l *.txt would display all .txt files in the current directory, with their permissions, such as:

```
-rw-rw-rw- 1 ownerid groupid 1428 Apr 2 19:42 myfile.txt
-rw-rw-r- 1 ownerid groupid 914 Sep 30 10:17 personnel.txt
```

The firs t 10 characters (or "switches") on each line contain the permissions for that particular file, directory, folder or other object. For an explanation of what the characters mean, see the question, When I look at a list of files and directories, the names are followed by something *like -rw-rw-r— or lrwxrwxrwx. What does this mean?*

If you are using the K File Manager (KFM) under KDE (just click on the Panel icon that looks like a house in front of a manila folder), simply select View, then Long View, from the pull down menu to display all objects with their permissions. Gnome users will have to use the command line method.

Q.85. How do I change permissions?

Ans. If you are the system administrator (root), or the owner of a file, directory or other object, you have the authority to grant or remove permissions for other users and groups, using the ***chmod command. The operative symbols to remember are u for User (owner), g for Group, o*** for Other, a for All, r for Read, w for Write and x for eXecute. Precede the r/w/x characters with a plus sign (+) to add permission or a minus sign(-) to deny permission.

For example, to grant Read/Write access for the file myfile.txt to Groups and Others, use the command: chmod go+rw myfile.txt. To remove Write and eXecute access from all accounts, use: chmod ugo-wx myfile.txt or chmod a-wx myfile.txt. (Note that a, for All, is a convenient shortcut for ugo.)

Communications : These questions relate to configuring or using communications features of Linux .

Q.86. How do I set up my system for dial-up Internet access?

Ans. This is a multipart process. Before you begin you should contact your Internet Service Provider (ISP). There is some information you will need to provide to Linux that only your ISP can provide. (A technician who is convers ant with Linux can probably step you through the entire process, but because you may not find a Linux ex pert at your ISP, here are the steps necessary to do it yourself.)

- If you are using KDE, use the Kppp utility to start the configuration process.

  Log in as the root operator, then use your favorite text editor to open the **/etc/ppp/ options file. If you see a statement with the option "lock" you will need to** "comment it out" to disable it. (Kppp handles locking the connection and the hardware, so Lock is redundant.) To disable the lock option, insert a # symbol in front of it, such as: #lock. Save the file and exit the editor. Log out as root and log back in as yourself.

  To start Kppp, go to the KDE Application Starter menu (the big K icon on the Panel) and select Internet, then Kppp.

  When the Kppp Configuration panel appears, press the Setup button to create a new ISP connection record.

From the Accounts tab, press the New pushbutton.

When the Edit Account panel appears, you will be in the Dial tab. Enter whatever you want to use as the ISP name where it says Connection Name. Enter the dial-up connection phone number in the next field (including area code, or anything needed to dial out of the switchboard, such as 9,). The default authentication type is PAP. Use that unless your ISP tells you otherwise. (The remaining fields are for experienced users. Unless you have a reason to enter information in those fields, leave them blank .)

Click on the IP tab. When that panel appears either choose Dynamic IP Address, or enter a static address. Your ISP will have to tell you which option to use, but generally the ISP will assign a temporary IP address each time you connect, so most likely the ISP will tell you to select Dynamic IP Address. If you are setting up a connection inside a firewall, you will probably be us ing a static address. The administrator in charge of the firewall will assign you a static address and tell you the subnet mask to enter in the appropriate fields. Use the "Auto-configure hostname from this IP" chec kbox if you are connecting to a DHCP server or a Bootp server. (Your network administrator will tell you if you need this.)

If you are connecting from home, you can skip the DNA and Gateway tabs. If you are connecting in an office environment, your network administrator will tell you if you need to enter anything in those tabs. Unless told otherwise, you should not need to enter anything in the Login Script tab either. If you live outside the United States or Canada and your ISP charges by bytes transferred, us e the Accounting tab to select the country, the company and whether you are charged by incoming data, outgoing, or both; otherwise ignore it. Click the OK button to close the Edit *Account panel and return to the Kppp Configuration panel.*

Click on the Device tab to set up your modem connection. If the modem was set up correctly during installation, you should have to select only the port to use. COM1 through COM4 are mapped respectively to /dev/ttyS0 through /dev/ttyS3. Use the one that corresponds to the port to which your modem is connected. An internal modem may be mounted as / dev/modem. For Flow Control, use CRTSCTS to enable hardware flow control. For Line Termination, choose CR/LF. For a 14.4Kbps modem, choose a connection speed of 57,600bps. For faster modems, select 115,200bps. If you would like to prevent a second program from disconnecting one already using the modem, check the Use Lock File box.

In the Modem tab, you will configure the modem. Use the Busy Wait box to specify how long to wait for a busy modem or dirty phone line before disconnecting. The *Modem volume slider lets you decide how loud to make the connection tones. Use the Modem Commands button to set the default control codes for certain functions. The Query Modem button can be used to test the modem, and the Terminal pushbutton* pops up a mini-terminal for entering control codes manually.

Click the OK button at the bottom of the panel to close the KPPP Configuration panel and return to the KPPP panel.

In the Connect to field, select the ISP for which you just created a record. Then enter the Logon ID and Password you were assigned. Make sure the Show Log Window box is checked. This will let you see the progress of the logon process, which is helpful in case there are problems logging on. Then press the Connect button to dial the ISP. If you are unable to connect, verify all of the information you entered in previous steps. If you do not find anything wrong, contact the ISP or network administrator for further assistance.

- If you are using Gnome rather than KDE, there is nothing packaged with the standard Linux distributions, but there is a downloadable tool you might try called Gnome Dialup. As of this writing, it is still under development but worth a try if you are looking for a graphical dial up access setup tool for Gnome. Gnome Dialup c an be found at *http:// tor-pw1.netcom.ca/~phantom/gnome_dialup/index.html. Other such tools may exist* as well.
- From the command line, you c an use the pppd command to start the PPP daemon. This approach is not for the faint of heart, however. There are many parameters that need to be set via this command, so unless you are quite knowledgeable about Point to Point Protocol setup or know someone who is, this probably is not the best approach for you. For more on the pppd command, read the instructions at: info pppd.

If you need help getting connected to the Internet via your office network, contact your help desk. If you are trying to set up Linux for DSL or cable modem (non-dialup) access at home, contact your DSL or cable modem provider for assistance (and hope someone there has ever heard of Linux).

Miscellaneous : This section contains questions or problems that did not fit into any of the other categories.

Q.87. What is the difference between Linux and UNIX?

Ans. UNIX began as a proprietary operating system developed by Bell Laboratories in the 1960s. It eventually spawned a number of mutually incompatible commercial versions from such companies as Apple (Mac OS X), Digital (Digital UNIX), Hewlett-Packard (HP-UX), IBM (AIX®), NeXT (NeXTSTEP) and others.

Linux is an attempt to create a nonproprietary operating system for the masses. It is a free, open source, UNIX-like operating system, originally begun by Linus Torvalds in 1991. "Linux" really refers to only the operating system kernel, or core. More than 200 people have contributed to the development of the Linux kernel. The rest of a Linux "distribution" package consists of various utilities, device drivers, applications, a user interface and other tools that generally can be compiled and run on other UNIX-based operating systems as well.

Q.88. Why does Linux seem so haphazard?

Ans. You may have heard the old saw that a c amel is a horse designed by committee, illustrating the difficulty of getting a number of people to agree on the design of anything, and the results of compromise. If anything, Linux illustrates a horse designed by complete anarchy, with a healthy helping of Darwinian natural selection thrown in for good measure. Thousands of programmers over the years have written applications, utilities, games, device drivers, user interfaces and other software for Linux. In the absence of a central authority to dictate a mandatory "look and feel" each programmer had his or her idea of what was the best design for software and developed the programs accordingly. Over time certain software became more popular than others, which in turn spawned "work -alikes" and "near-clones," taking features of the previous products and adding to them, or offering slightly different variations.

Eventually, companies such as Red Hat, Caldera, TurboLinux and SuSE took the Linux kernel and added what they considered the best available utilities, user interfaces, etc., to form a "distribution," or packaged deliverable. Because each distribution chose from among this programming chaos, it was inevitable that the various programs selected for a deliverable looked and worked differently from one another, and that each deliverable looked and, to some extent, worked differently from other deliverables. This is the reason for the inconsistencies between programs in a deliverable. The flip side is that because the distribution companies were free to choose from among all available programs, not just those that slavishly conformed to an arbitrary style, the deliverables were able to include the very best programs—even the ugly or nonconformist ones.

Q.89. Where can I download or buy Linux software?

Ans. There are quite a few sources for free or commercial Linux software if your local software store doesn't carry what you are looking for. Here are just a few of them:

For free software, try:

- DLR Fresh Archive (http://www.go.dlr.de/fresh/linux/src)
- Freshmeat (http://freshmeat.net; love the name!)
- Linux Apps (http://www.linuxapps.com)
- Linux Archives (http://home.linuxarchiv es.com/software.html)
- Linux Game Tome (http://www.happypenguin.org)
- Linux Games (http://www.linuxgames.com)
- Linux Software Encyclopedia (http://stommel.tamu.edu/~baum/linuxlist/linuxlist linuxlist.html)
- Linuxberg/Tucows Linux (http://linux.tucows.com)
- Slashdot (http://slashdot.org)

If you see some c ommercial software that you would like to buy, there are several Web sites in the United States with extensive Linux inventories:

- Indelible Blue (http://www.indelibleblue.com)
- Linux Mall (http://www.linux mall.com)

- The Linux Store (http://www.thelinuxstore.com)

In Australia, Linux users can go to:

- Everything Linux (http://www.everythinglinux.com.au)

Beyond merely software, thes e sites also carry Linux books, and hardware devices known to be Linux-friendly, should you need to add to or upgrade your system. Some of them even offer such things as caps, mugs, neckties and other Linux paraphernalia. If you are look ing for an extensive list of Linux books, try O'Reilly & Associates (http://linux.oreilly.c om).

Q.90. Where can I go for assistance on Linux?

Ans. There are a great number of Internet sources for Linux assistance, including vendor Web sites, independent Web sites and Usenet newsgroups. Some informational Web sites include:

- Caldera Linux support page (http://www.caldera.com/support).
- Debian Linux support page (http://www.valinux.com/services/support ?session_hash=7d7ccc8f28c0f42cddc98da037d7875a).
- Gnome Web site (http://www.gnome.org) — Home of the Gnome desktop environment.
- KDE Web site (http://www.kde.org) — Home of the KDE desk top environment.
- Linux at IBM (http://www-4.ibm.com/software/is/mp/linux) — The IBM Linux home page, containing news, software downloads and support options.
- Linux Documentation Project (http://www.linuxdoc.org) — A c ollection of Frequently Asked Questions (FAQs), how-to documents, manuals and online magazines.
- Linux Gazette (http://www.linuxgazette.com) — It includes articles, columns and even comic strips.
- Linux Guide (http://www.firstlinux.com/guide) — A glossary of Linux and communications terms.
- Linux Journal (http://www.linuxjournal.com) — It contains links to newsgroups, chat rooms, online manuals, FAQs, vendor support sites, local Linux user groups and more.
- Linux Man Pages (http://linux.ctyme.com) — A collection of the man (manual) instructions for various Linux commands and utilities.
- Linux Newbies (http://www.linuxnewbies.org) — Help files for "newbies" (those new to Linux).
- Red Hat Linux support page (http://www.redhat.com/apps /support).
- Slick Penguin (http://www.slickpenguin.com) — White papers , case studies and other Linux business-related implementation success stories.
- SuSE Linux support page (http://www.suse.com/us/support/index.html).
- TurboLinux support page (http://www.turbolinux.com/support).

The following newsgroups can provide a significant amount of user-to-user help:

- comp.os.linux.announce — Announcements of new products, updates, bug fixes, etc.
- comp.os.linux.hardware — Support on hardware issues, including compatibility, configuration, device drivers and product evaluations.
- comp.os.linux.misc — A catc h-all for whatever doesn't fit in one of the other groups.
- comp.os.linux.networking — For questions about networking hardware, software and configurations.
- comp.os.linux.setup — How to install and configure Linux and add-on produc ts.
- comp.os.linux.x — Installing, configuring and using the X Window System under Linux.

There are other Linux newsgroups provided for other topics, such as vendor-specific questions (alt.os.linux.caldera, linux.debian.user or redhat.rpm.general, for example) and those for programmers.

For current news about Linux and article archives, visit:

- CNET Linux Center (http://linux.cnet.com/?tag=st.ne.ni.refer.1491268) — A Linux-specific news site. It not only has news, but also link s to popular Linux downloads, product reviews, Linux events, Linux company stock quotes and other resources.

- ICE News (http://www.nikos.com/icenews/linux.html).
- Linux Planet (http://www.linuxplanet.com/linuxplanet).
- Linux Today (http://linuxtoday.com/index.html).
- Linux Weekly News (http://www.lwn.net).
- Linux World (http://www.linuxworld.com).
- Zdnet Linux Homepage

❐❐❐

INDEX

K...

L...

M...

N...

O...

P...

R...

S...

T...

U...

W...

❐❐❐

Win Prizes

ATTENTION READERS

Win Prizes

We request for your frank assessment regarding various aspects of the book as given below :

SYSTEM ADMINISTRATION

By: PANKAJ SHARMA

Please Fill up the Sheet in Neat and Clean Writing

(i) What topics of your syllabus, important for your examination point of view, are not covered in this book ?

(ii) What are the chapters wherein the treatment of the subject matter is not systematic or organised or updated ?

(iii) Have you come across misprints/mistakes/factual inaccuracies in the book? Please specify the chapters and the page numbers.

(iv) What other books on the same subject have you found/heard better than the present book? Please specify in terms of price and quality

(v) Further suggestions and comments for the improvement of this book.

(PS: If need be please attach a separate sheet to write your views)

OTHER DETAILS

(i) Who recommended you this book ? (Please tick (✓) in the box near the option relevant to you.

☐ Teacher ☐ Friends ☐ Bookseller

....

....

....

....

(ii) Name of the recommending teacher, his designation and address :

....

....

....

....

(iii) Name and address of the bookseller you purchased the book from :

....

....

....

....

(iv) Name and address of your institution (Please mention the University or Board as the case may be)

....

....

....

....

(v) Your name and complete postal address :

....

....

....

....

(vi) Course you are enrolled in :

....

....

....

....

The best assessment will be awarded each month: The award will be in the form of our publications as decided by the Editorial Board.

Please mail the filled up coupon at your earliest to :

S.K. KATARIA & SONS®
4760-61/23, Ansari Road,
Daryaganj, New Delhi-110002
☎: +91-11-23243489, 23269384
Email : katariabooks@yahoo.com
katariabook@gmail.com